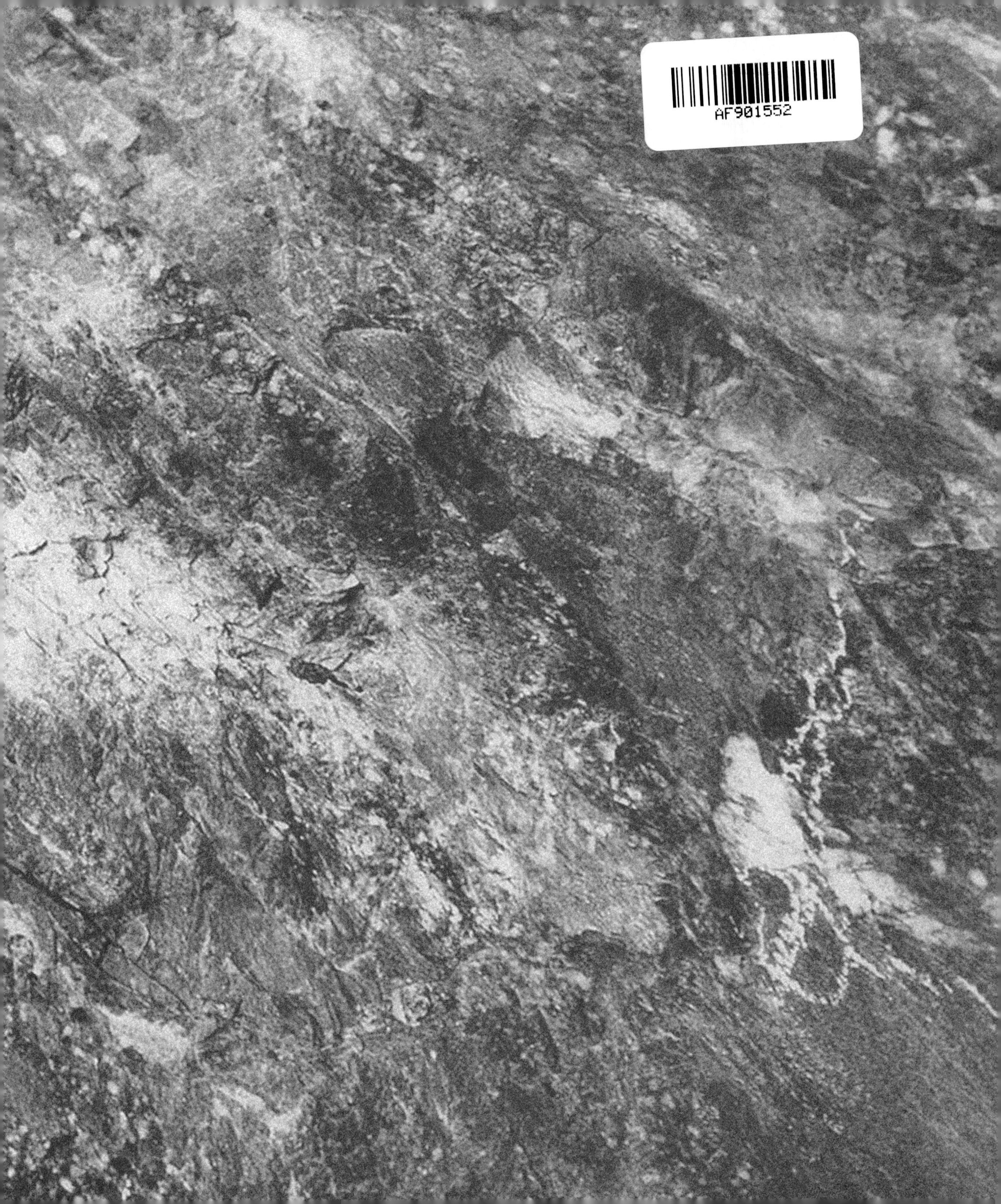

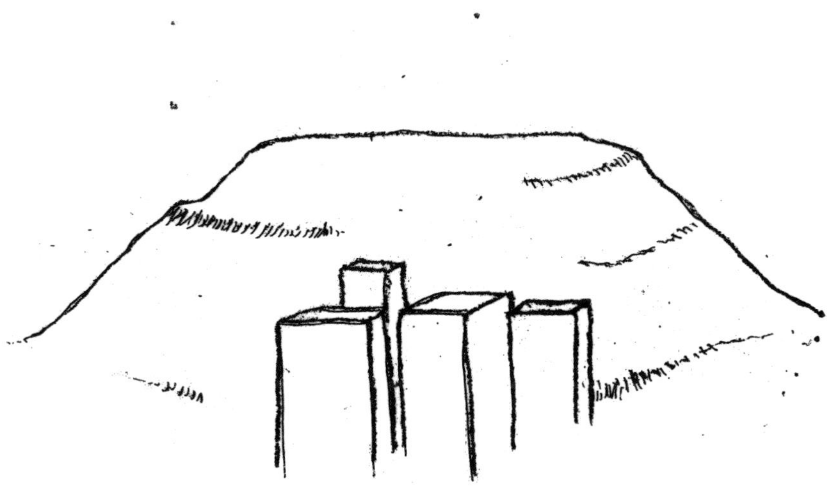

TALES FROM THE CITY OF GOLD

JASON LARKIN

KEHRER

N8
Soweto, Johannesburg, 2010

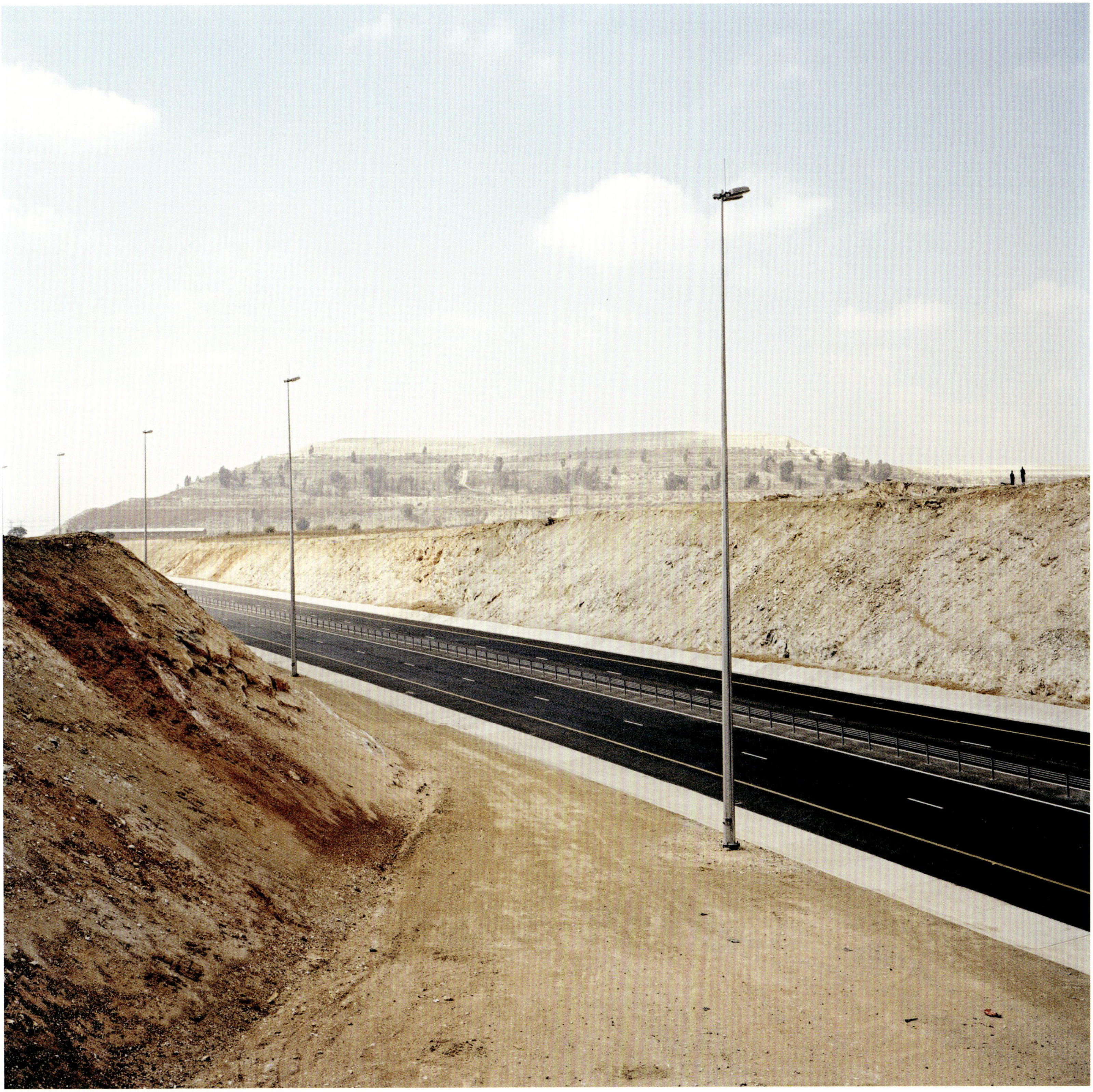

THE REEF & THE CITY

MARA KARDAS-NELSON

At first sight, it seems unlikely that Johannesburg's Main Reef Road could provoke nostalgia. But this otherwise unusual and unkempt highway, with its worn tar and its dimming yellow lines, acts as a gateway to hundreds of mines – both active and inactive – that dot the metropolis. It was along this road that its namesake gold reef was discovered in 1886, a discovery which turned a small, desolate farming community into an international mining mecca within a matter of years.

This transformation may belong to the past, but its impact is embedded into the fabric of modern Johannesburg, and the dusty face of Main Reef Road ensures that it cannot be forgotten. Travel east to west on the 160 kilometre-long stretch of highway, which leads from small, largely decrepit farming and mining towns and straight into the booming city centre, and you will be met at regular intervals with big golden-glow heaps that line the embankment.

These are the mine dumps, also known as tailings. They are everywhere, a constant reminder of a city's past and – welcome or not – an entire country's uncertain future. After the mines came the tailings. What comes after that?

Johannesburg is, quite literally, a city built on gold. It sits atop the Witwatersrand Basin, which once housed one of the world's largest deposits of the precious metal. Although most of the basin's once extensive deposits are now gone, everything urban stems from this mining past: the grey vertical faces of the Anglo American and Chamber of Mines buildings, which are landmarks in the city centre; the now-booming banking sector, which was originally financed by and made to support the mining industry; the busy M2 highway, which runs from the city's centre to Soweto and skirts crusted mine dumps. Mining is the city's ecosystem: everything can be traced back to the subterranean riches of old.

The sector's most obvious mark can be seen in Johannesburg's physical and racial landscapes. Its layout, which dictates what the city looks like, and how its people interact, is built soundly on the policies and practices instituted through its mining past. When diamonds, and then gold were first discovered, mine managers, unable to afford European labour, which was in high demand and short supply, instead relied on cheap, unskilled, and abundant African labour. Bosses wanted strict control over this labour force, and so established the now-infamous male-only dormitories, otherwise known as hostels, as well as a pass system to identify and restrict the movement of African workers. Those not working in the mines but still hoping to find employment in a rapidly expanding Johannesburg found shelter in the growing settlement's central district, located—as it still is—away from the immediate dirt and dust of the mines, but still close to work and amenities.

But the presence of African workers in the city offended white sentiment, and was also seen to counter the mines' need for ready, accessible labour. In the 1930s—years before the National Party, responsible for formally instituting apartheid, came to power in 1948—blacks were moved to settlements closer to the mines in order to keep the city centre 'white,' and labour close. Their new homes lay on unfavourable land to the south, where the Highveld winds incessantly blew mine dust—waste from a prosperous industry—over everything. Whites, meanwhile, took up occupancy in the northern part of the city, where a green belt of planted trees and parks kept the dust at bay. Today, the northern suburbs of Johannesburg are still largely white neighbourhoods, their affluence stark in contrast with the poverty found in black townships, which lie only a few kilometres away.

Further evidence of the city's mining history is seen in the environmental and industrial remnants that this thriving industry has left behind. Billions of tons of mine dust are housed in hundreds of massive dumps, which freckle the entirety of Johannesburg. The dust is finer than sea sand, the result of rock worn to nothing through decades of pulverising extraction. Along with polluted water, it is the most hazardous physical byproduct of the mining process.

In the early days of the gold rush, mine waste was deposited onto ever-growing dumps that sometimes sat within metres of an active mine, and equally close to encampments and hostels. Johannesburg was far from a fully-fledged city then, but rather a small, struggling community of hopeful prospectors: there were no roads and skyscrapers and millions of inhabitants to speak of, but instead only a migratory, seemingly expendable population. Gold was for now; city planning was an afterthought. Development and infrastructure followed piecemeal, and primarily catered for the needs of the industry. The mines, marked by the piles of dust that steadily grew beside them, were built haphazardly wherever the metal was struck.

But unexpectedly, and in spite of this tremendous lack of planning and foresight, Johannesburg didn't experience the boom-bust phenomenon that other mining towns did. The industry—and the city that sprung from it—showed unexpected longevity. South Africa's unrivalled gold deposits allowed the country to act as the world's largest producer until the mid-1990s. Underground wealth rapidly transformed the landscape from a speckled strip of defiant Afrikaner farms and African settlements, into Africa's biggest city.

And while the sector remained omnipresent, the city was not limited to mining. Its profits fueled the manufacturing and banking industries, both key to economic growth. In defiance of the steady depletion of gold deposits, the city continued to flourish, with industry spurring industry, and drawing labour from the region, and business from around the globe.

While the city remained, now a mecca unto itself, the inevitable gold crash finally happened, a result of fluctuating prices and the need to search for ever-deeper reserves. When gold prices hit a final dip in a series of lows in the 1990s, many of the last remaining industry giants fled the country, abandoning their operations, selling their assets, and dissolving into a series of smaller, often overlapping, companies.

But the dumps remained. They now act as a constant skeletal reminder of the past and as the industrial and environmental building blocks for today and tomorrow. Once just an afterthought, the dumps are now central to the city's physical and social identity.

After the mines, the industry's ghosts persisted. Even with the fall of apartheid and the slow demise of South Africa's gold industry, a pattern of segregated living shaped by the 19th century desires of the mining giants continues today. Many of the country's poorest citizens—mostly black South Africans and other Africans—continue to reside close to mining sites, either in the city's many townships, or in sprawling and proliferating informal settlements.

But there is one major difference between then and now: the townships and settlements of today are situated not near operational mines, but instead near mine waste. Informal settlements dot the bases of the dumps; some are even perched precariously atop. They house some of South Africa's most vulnerable, including many recent immigrants who have come to the city in search of work, following a cycle of migration stemming from economic push and pull factors that has played out for over a century. Today, nearly two million people live on or near six billion tons of mine waste.

Because of the mine dumps' close proximity to a major city, and South Africa's unique political history, Johannesburg's people and environment are arguably more vulnerable to the effects of mine waste than cities elsewhere. Under apartheid, the South African state supported mining houses, not only by creating conditions that suited their insatiable need for labour, but also by requiring little environmental accountability. The country's laws have changed significantly under democratic dispensation: they now require more stringent environmental protection, and place greater environmental and social responsibility onto mine owners. But a significant shortcoming remains: the government is not able to seek economic or restitutive redress from mining companies that have abandoned their tailings. Instead, old mining sites are either the financial and environmental responsibility of the government, or of any mining companies that buys them. As such, finger pointing is the norm, with the government claiming that operational mines must be held accountable for all liability, new or old, and mining houses saying that taking on the historic liability of their predecessors is unfair, and will lead to reduced investment in the country. Both insist the other should be responsible for cleaning up environmental waste.

And, with nearly 400 abandoned sites throughout the greater Johannesburg area, there is no shortage of waste to be dealt with. Environmental activists, community members, academics and the government are concerned about the effect that the dumps may be having on the city's ecology, and human and animal health. Many of the metals in the dumps—including uranium, lead and arsenic—are radioactive or toxic. While they are all naturally occurring, the mining extraction process brings them to the surface much more quickly than would occur through natural erosion, and in smaller, more 'digestible' quantities. Significant exposure may result in severe health

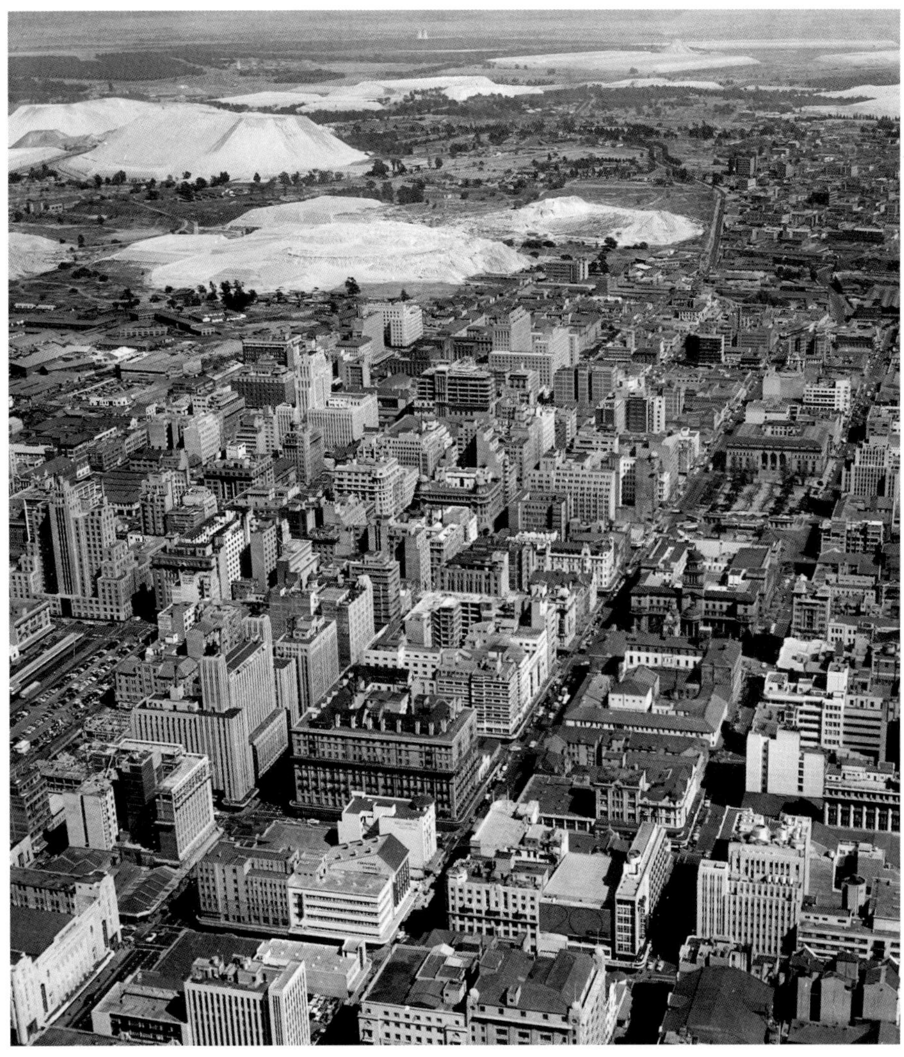

DOWNTOWN JOHANNESBURG
1960 (P.G. Higgins)

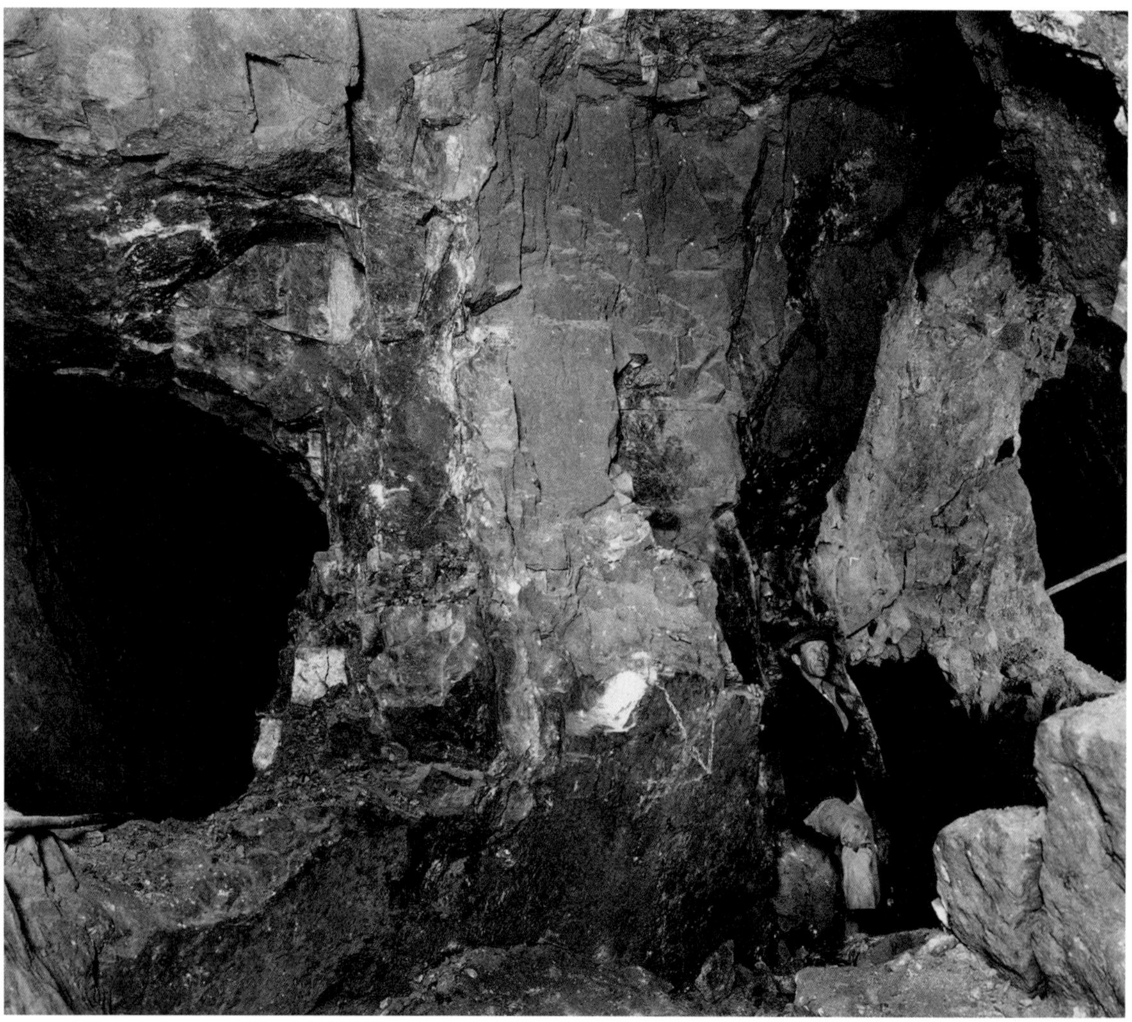

GOLD MINE SHAFT
c. 1910 (Unknown)
© Museum Africa

difficulties. Uranium generates the greatest fear: the most zealous refer to the potential threat from the country's mine waste as 'South Africa's Chernobyl'. While intense exposure to high levels of the element poses health risks, such as negative effects on the kidney, brain, liver, heart and endocrine system, the impact of long-term, low-level exposure is still unknown.

But there is now a newfound economic incentive for companies to care for and utilise the dumps. A spike in gold and uranium prices at the start of the new millennium caused increasingly desperate mining companies—often small, South African houses, as opposed to the American and European-owned giants of the past—to begin looking to the mine dust for new wealth. Technological advances now make it possible for the dumps to be re-mined for any precious metals still left in their mounds. It is estimated that since 2005, 40% of the country's tailings dams have disappeared. After the mines came the dumps, and after the dumps came new mines.

Visitors to Johannesburg often remark on the number of tailings, and their close proximity to everything: homes, schools, jobs, finance. Fly over the city, and you see them pockmark the landscape. On some, straggly nets sit atop to keep the dust down on windy days. On others, little green spots poke determinedly out of the sides and tops of mountains, plucky attempts at keeping erosion at bay. Others still have huge chunks bitten out of them, a sign of the mining sector's newfound resurgence and money being made out of waste.

But for those living in Johannesburg, they have become so profoundly integral to the landscape that they provoke little response. Few people seem to even notice them anymore: they are sometimes fondly referred to as "mountains," but most inhabitants see them as hosting barely any significance at all beyond acting as a place by which to know whether to take a left or a right off of one of the city's many roads that line its mining past.

How on earth, any visitor to Johannesburg may wonder, can a city live—grow, build, thrive—in the midst of these mountains of waste? How can there be so many dumps at the heart of things? And how can the dumps be so pervasive, and yet so invisible in the consciousness of ordinary citizens?

Perhaps the city-dweller's disinterest is a reflection of how South Africans have learned to cope with a city, and a country, which hosts such a complicated story, one that is as much about poverty, racism, inequality, crime and environmental degradation as it is about reconciliation, the resilience of people, and the beauty and strangeness of the nation's vista. After the mines came the dumps, and with them uncomfortable questions about what kind of country South Africa aims to be.

METRORAIL
Industria, Johannesburg, 2010

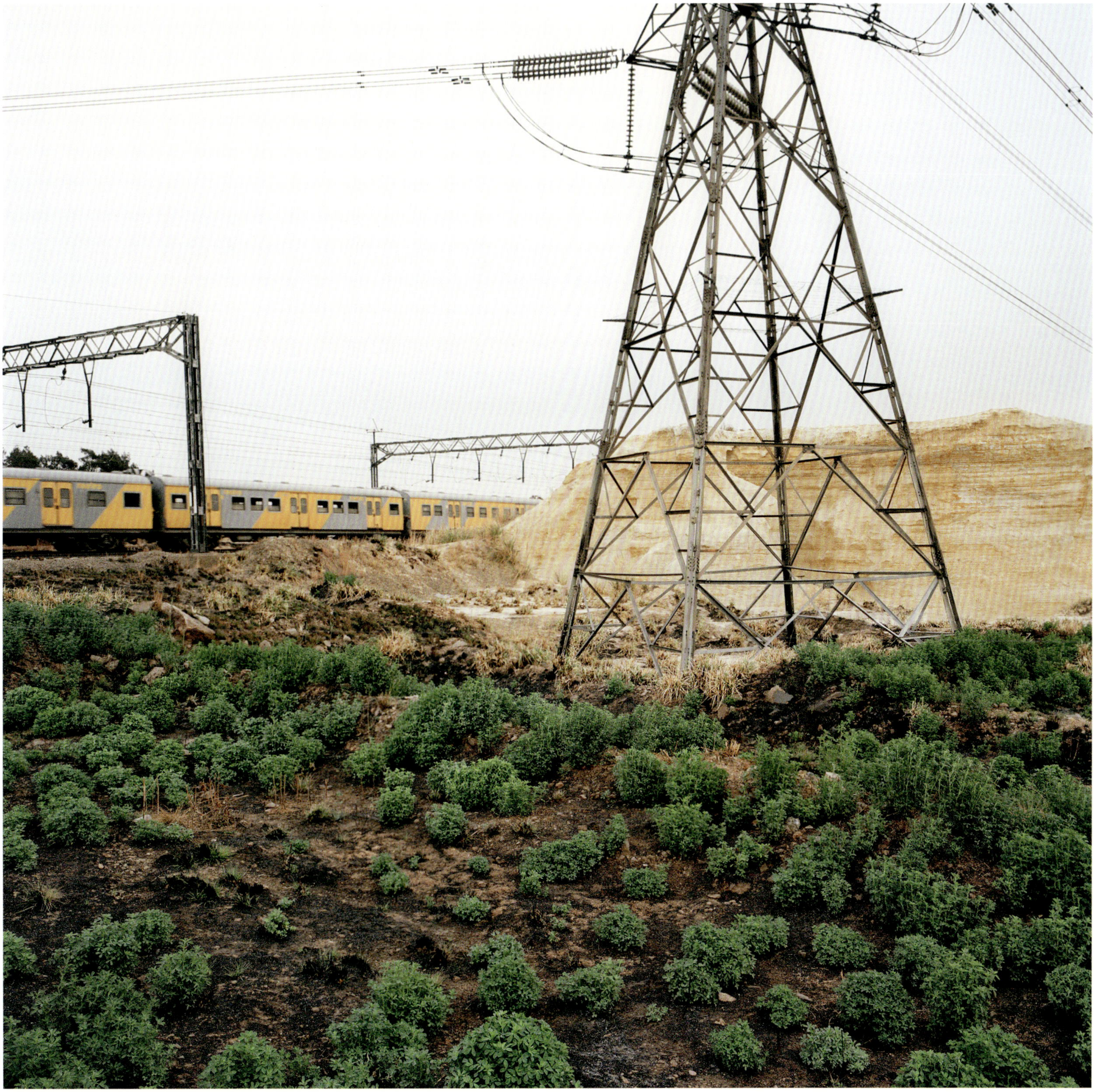

RDP HOUSING
Riverlea, Johannesburg, 2012

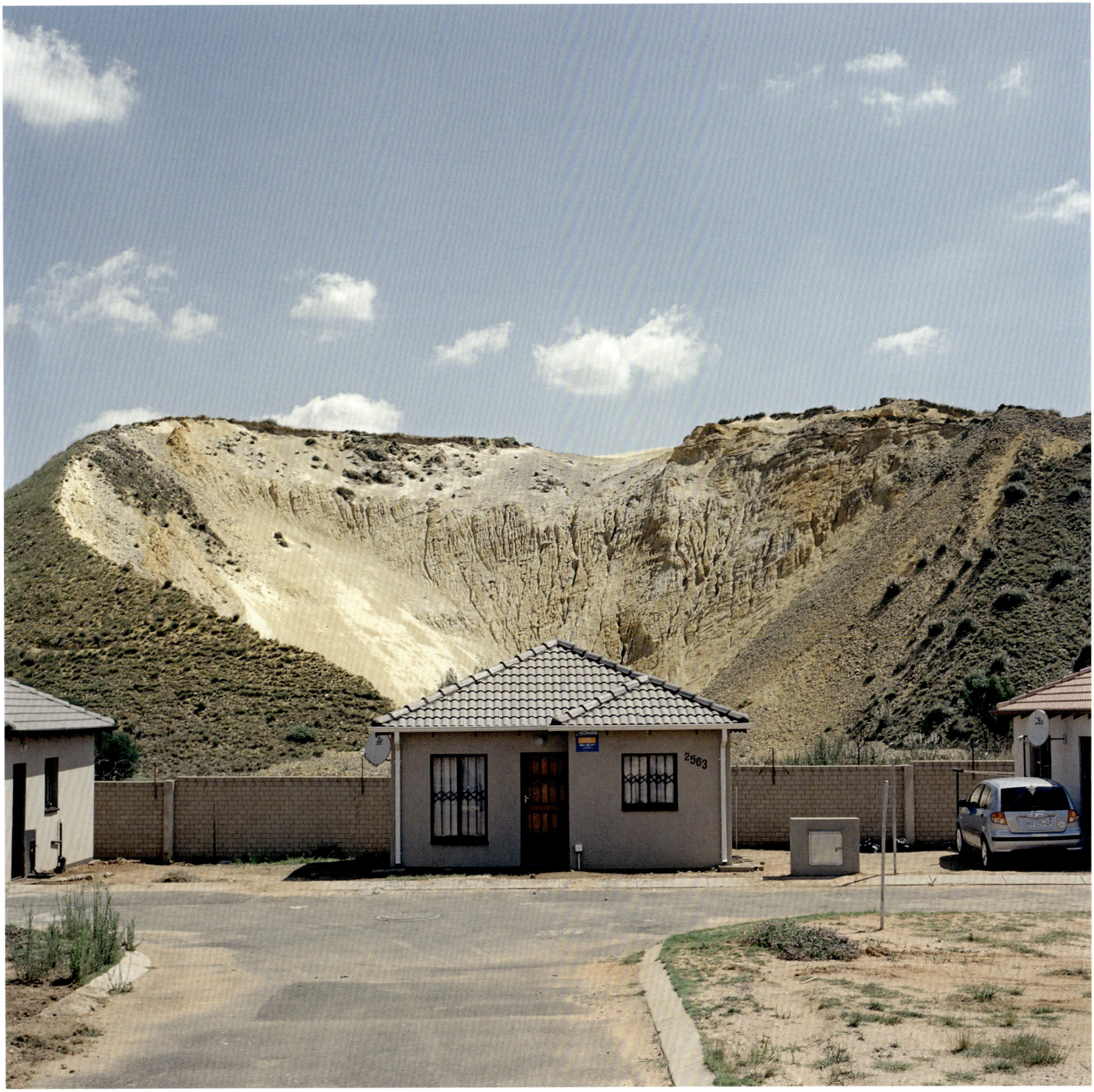
2563

COLOURED GIRLS
Riverlea, Johannesburg, 2012

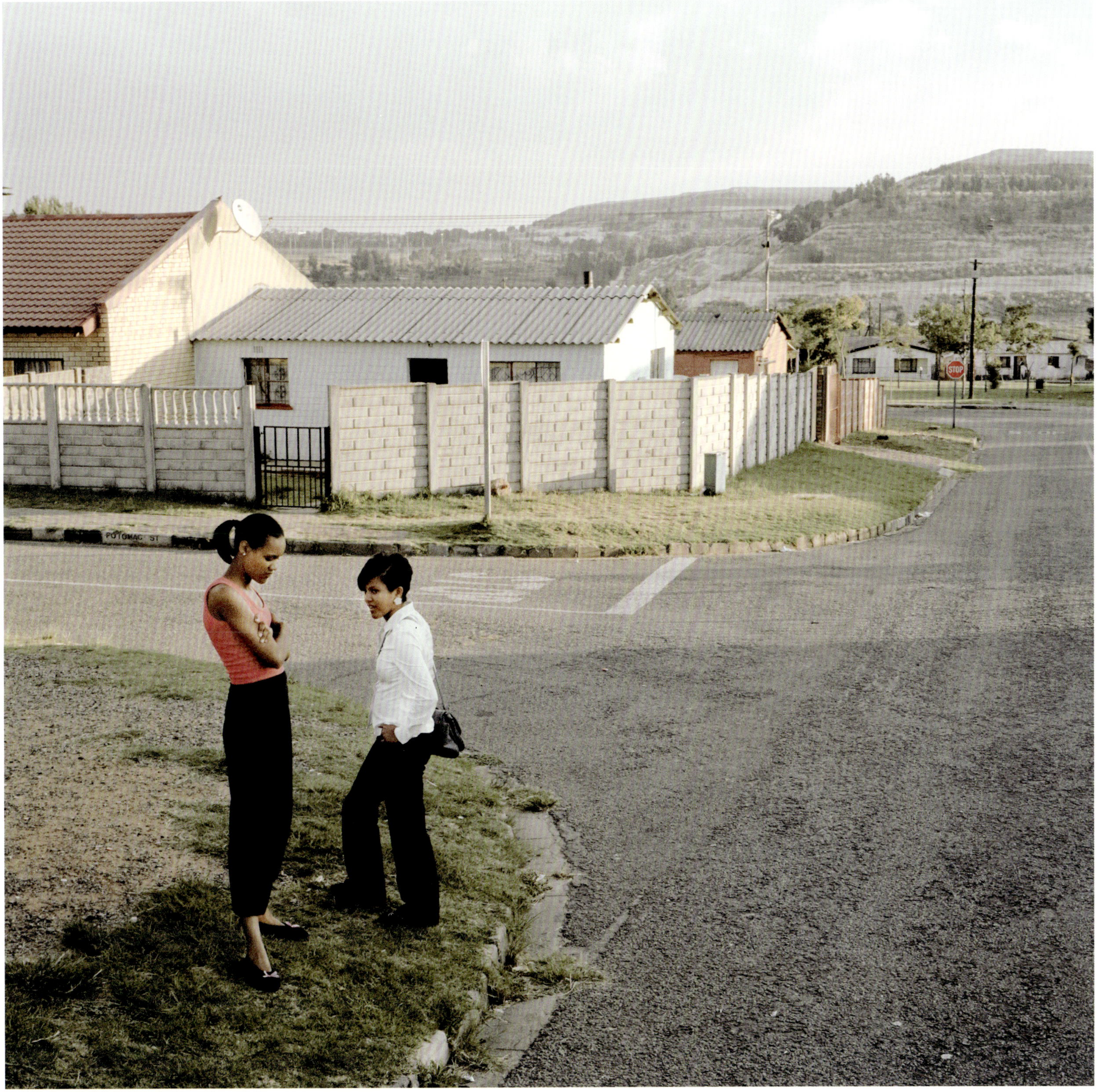
POTOMAC ST
STOP

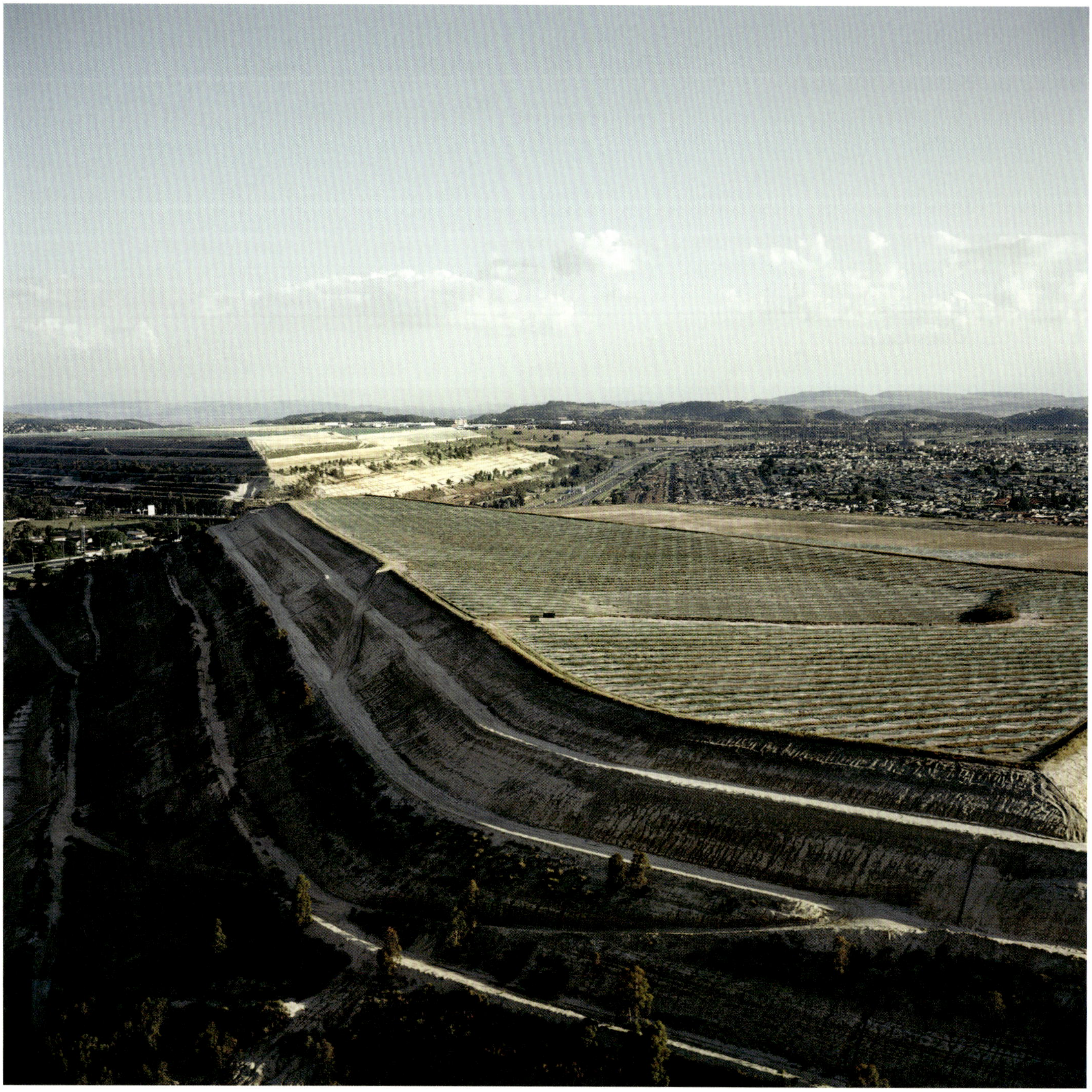

DIEPKLOOF
Soweto, Johannesburg, 2012

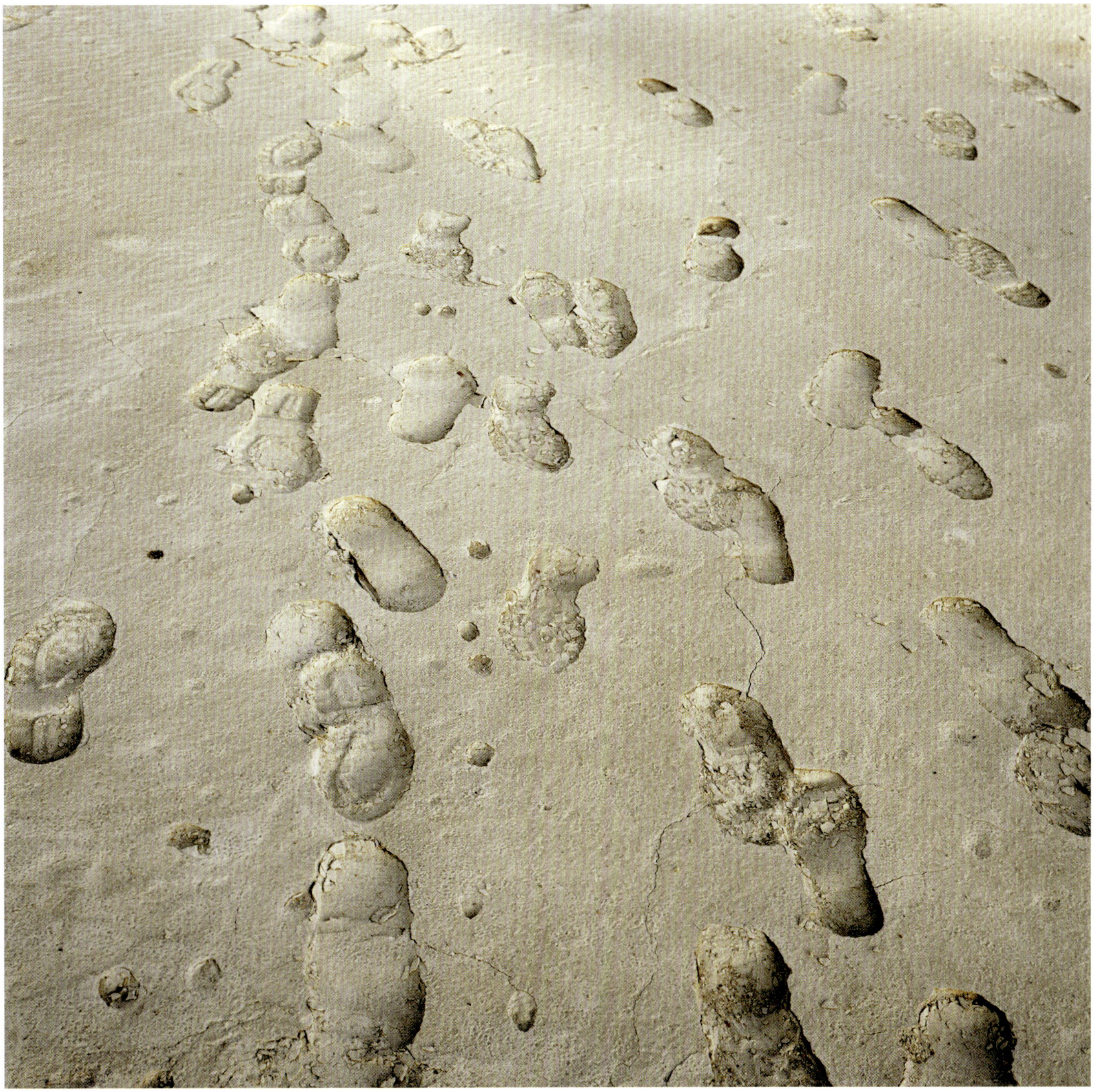

FOOTSTEPS
Randfontein, Johannesburg, 2010

DANIEL AND THE HUNTING DOGS
Selby, Johannesburg, 2012

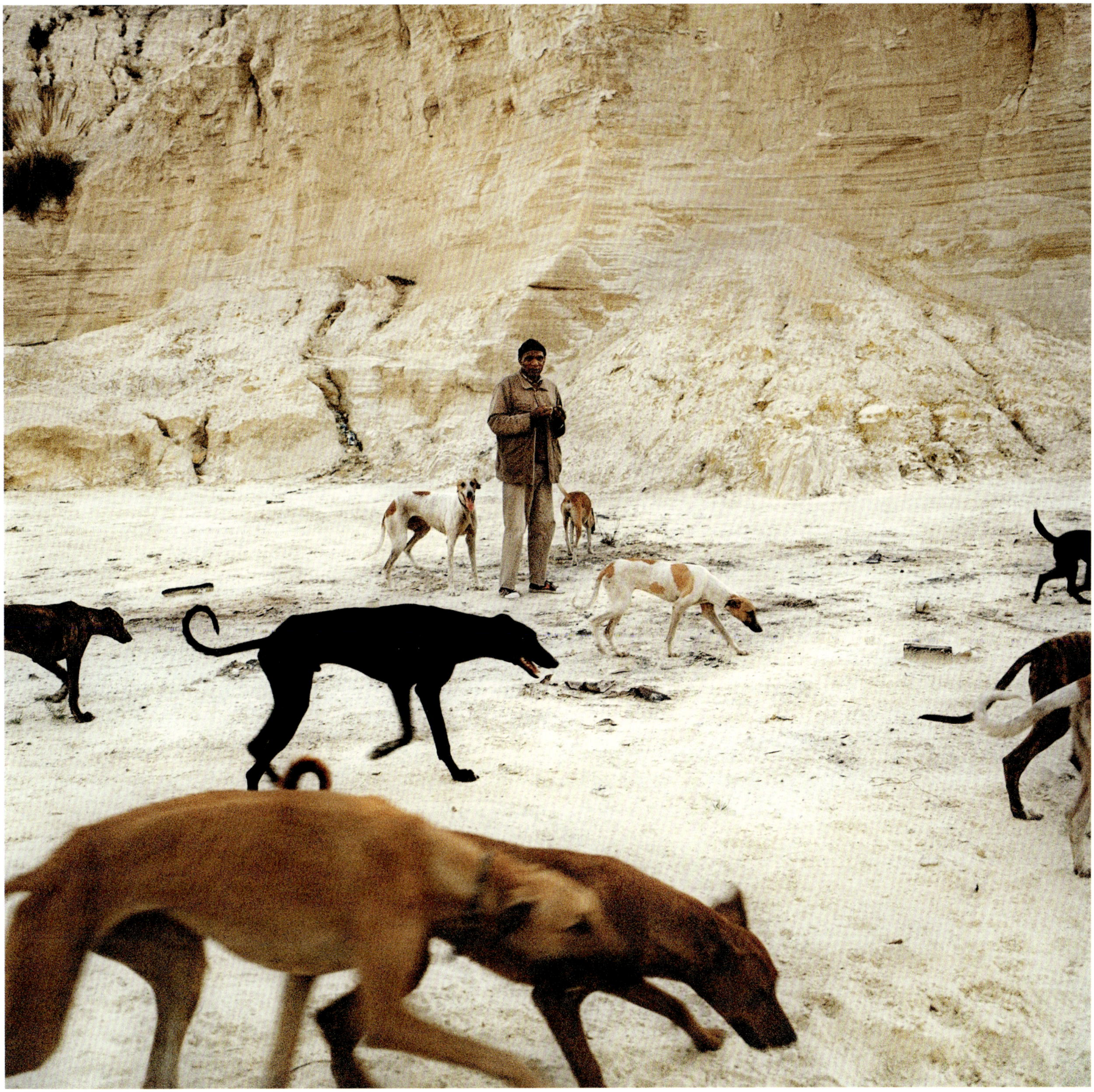

HOME TO PEOPLE AND HUNTING DOGS
Selby, Johannesburg, 2012

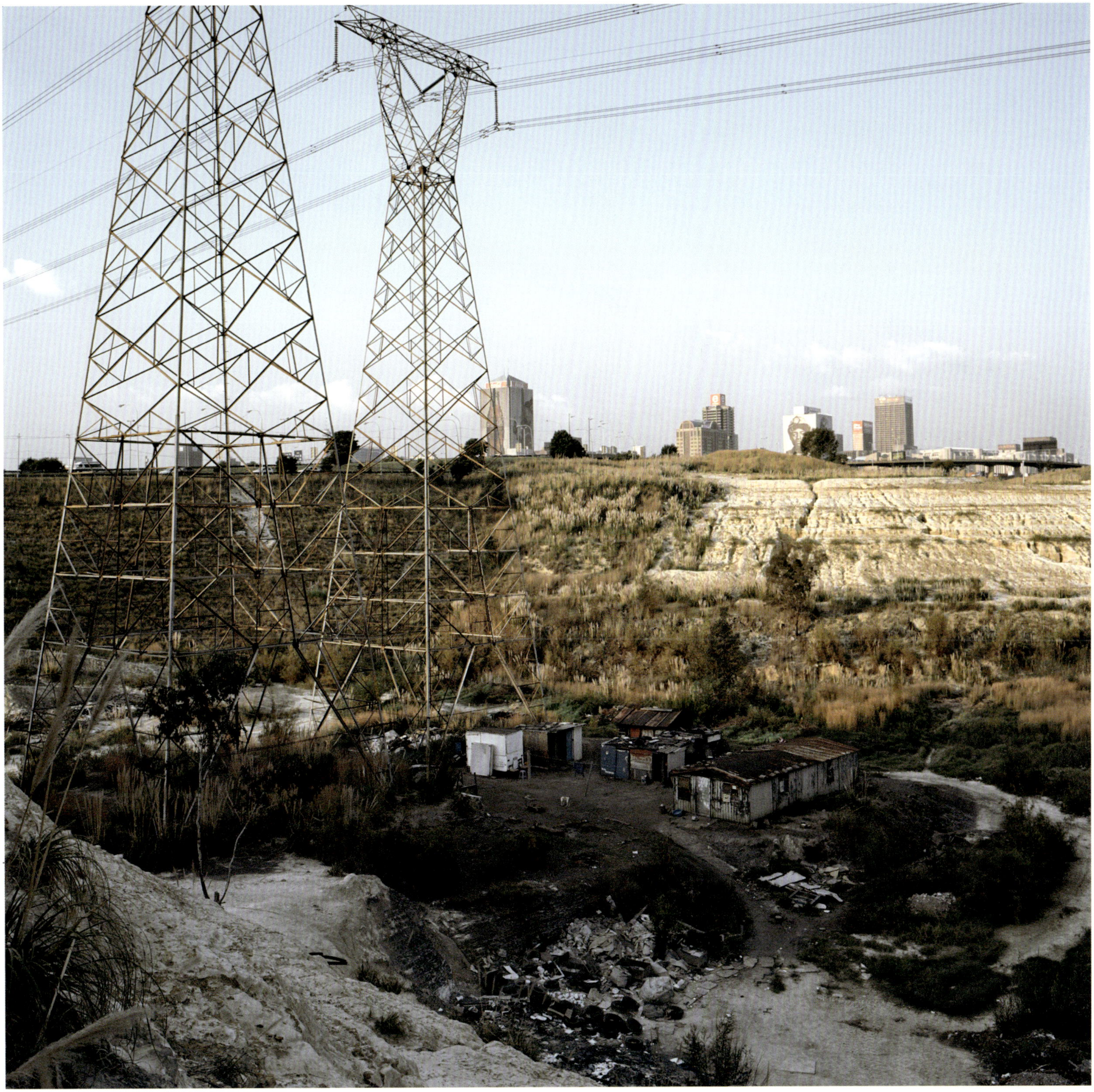

SPORTS PITCH, CENTRAL JOHANNESBURG COLLEGE
Diepkloof, Johannesburg, 2010

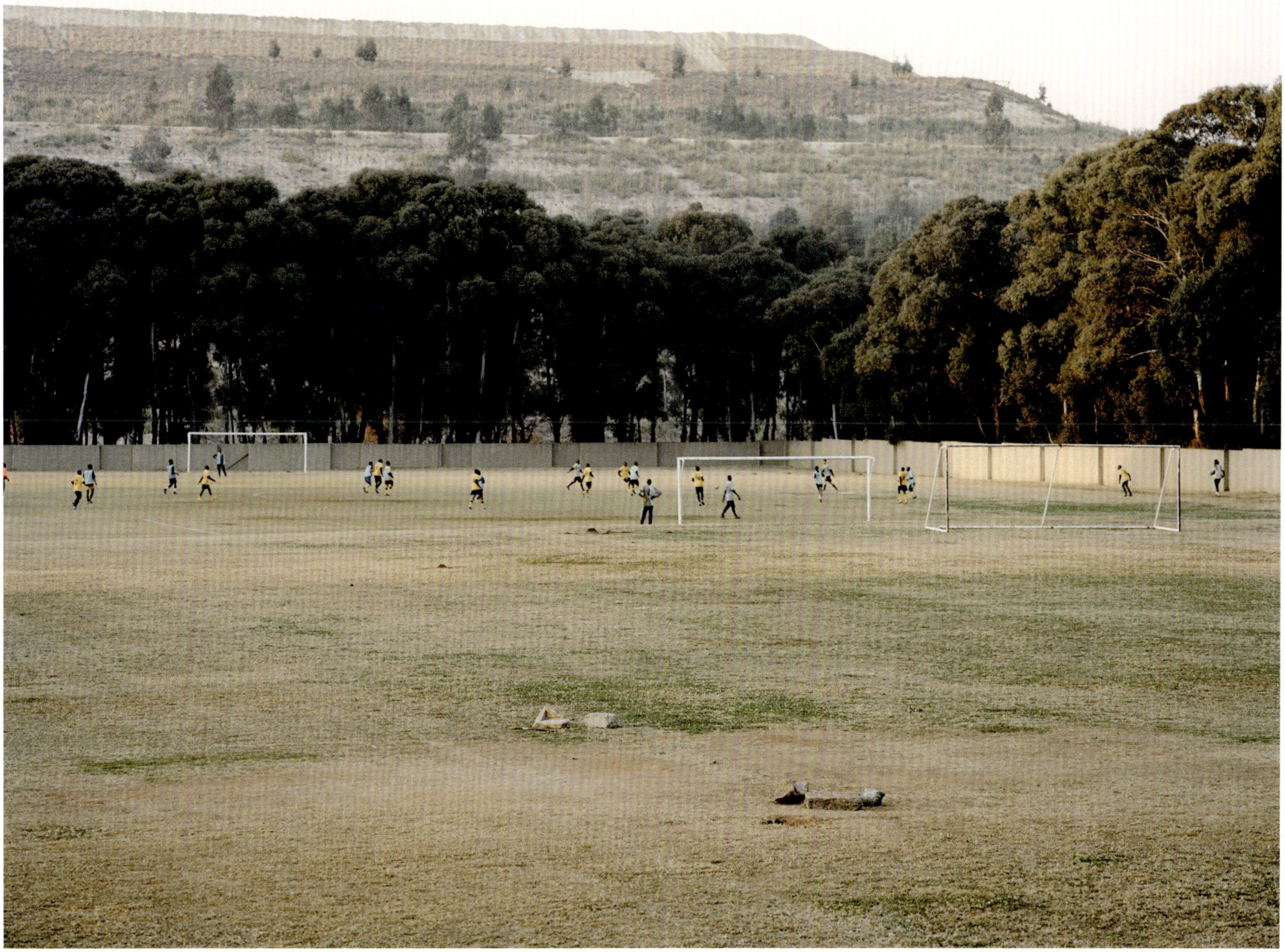

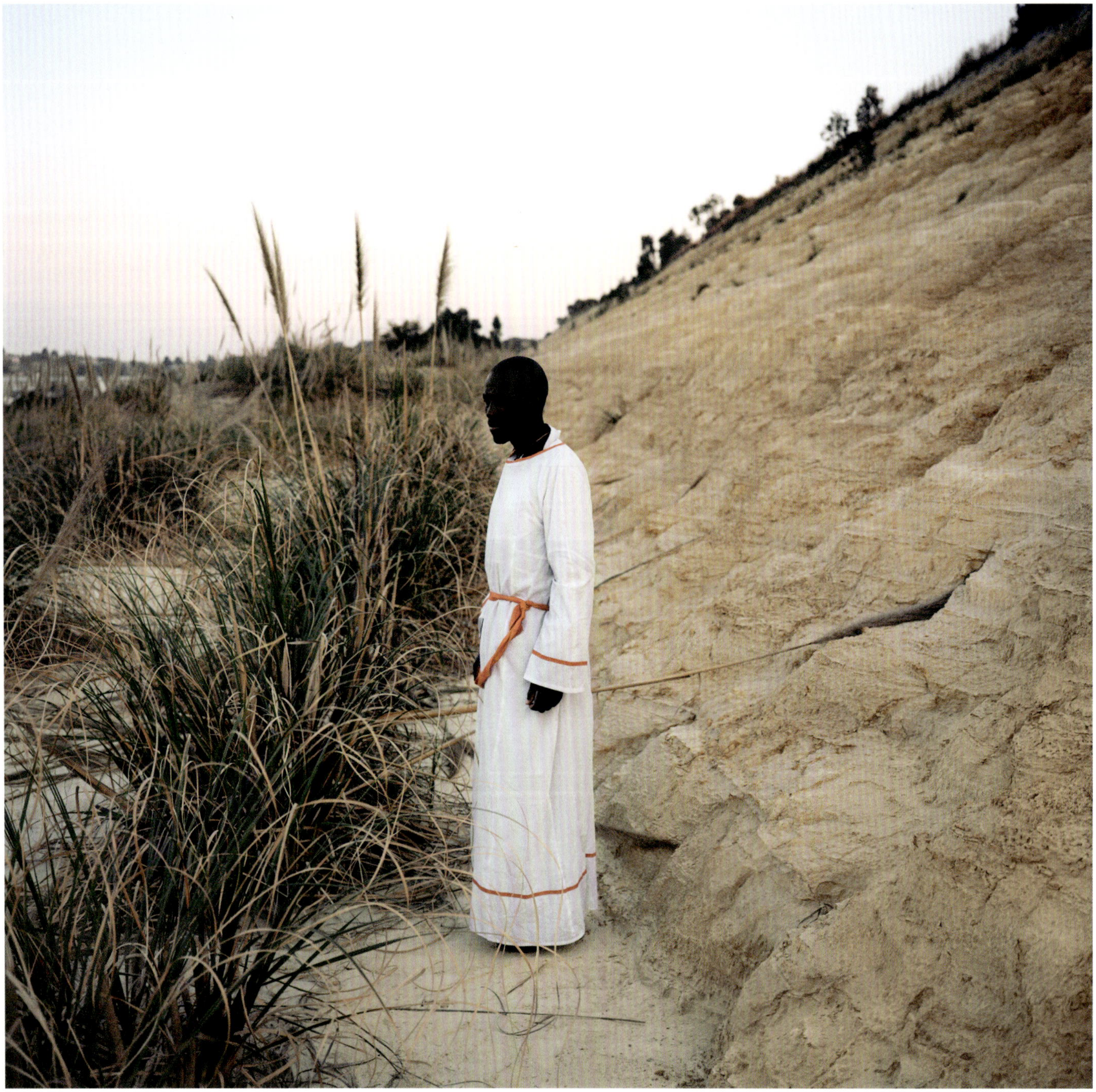

MIKE DWJELA, AFTER HIS EVENING PRAYER
Boysens, Johannesburg, 2010

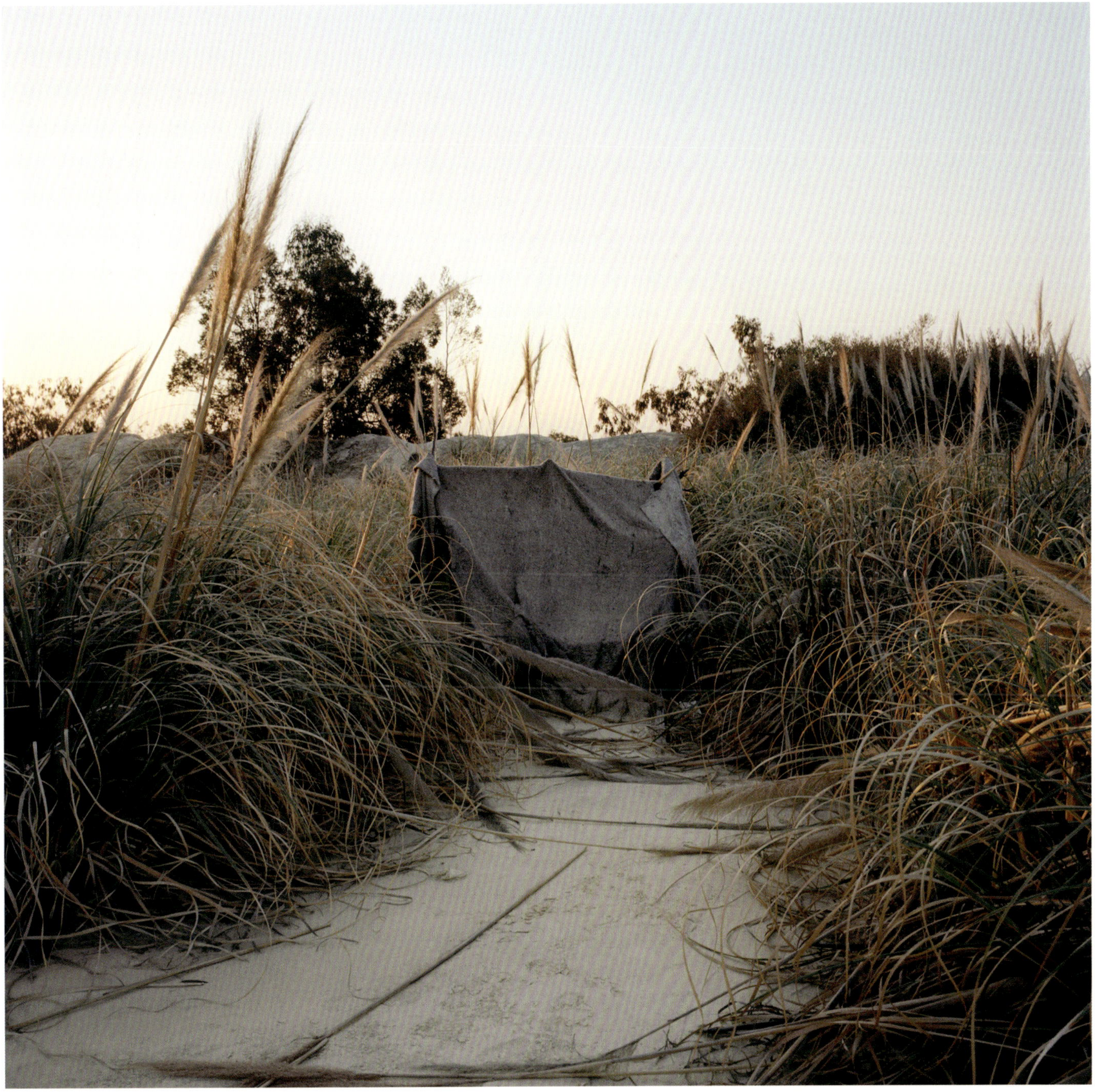

A HAND-MADE HOME
Boysens, Johannesburg, 2010

‘TOP STAR’
Boysens, Johannesburg, 2010

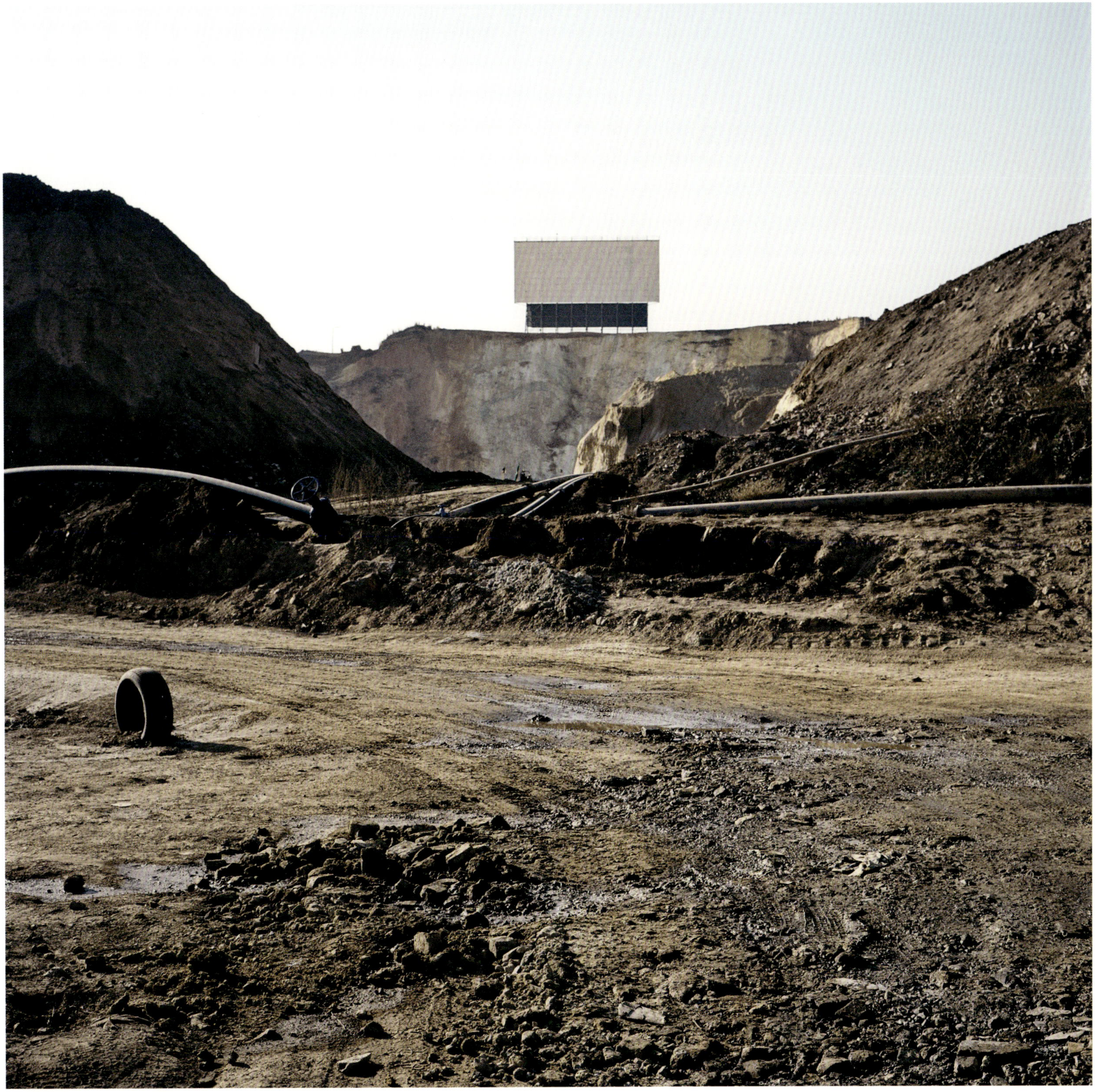

PLAYGROUND BUILT ON OLD DUMP
Riverlea, Johannesburg, 2012

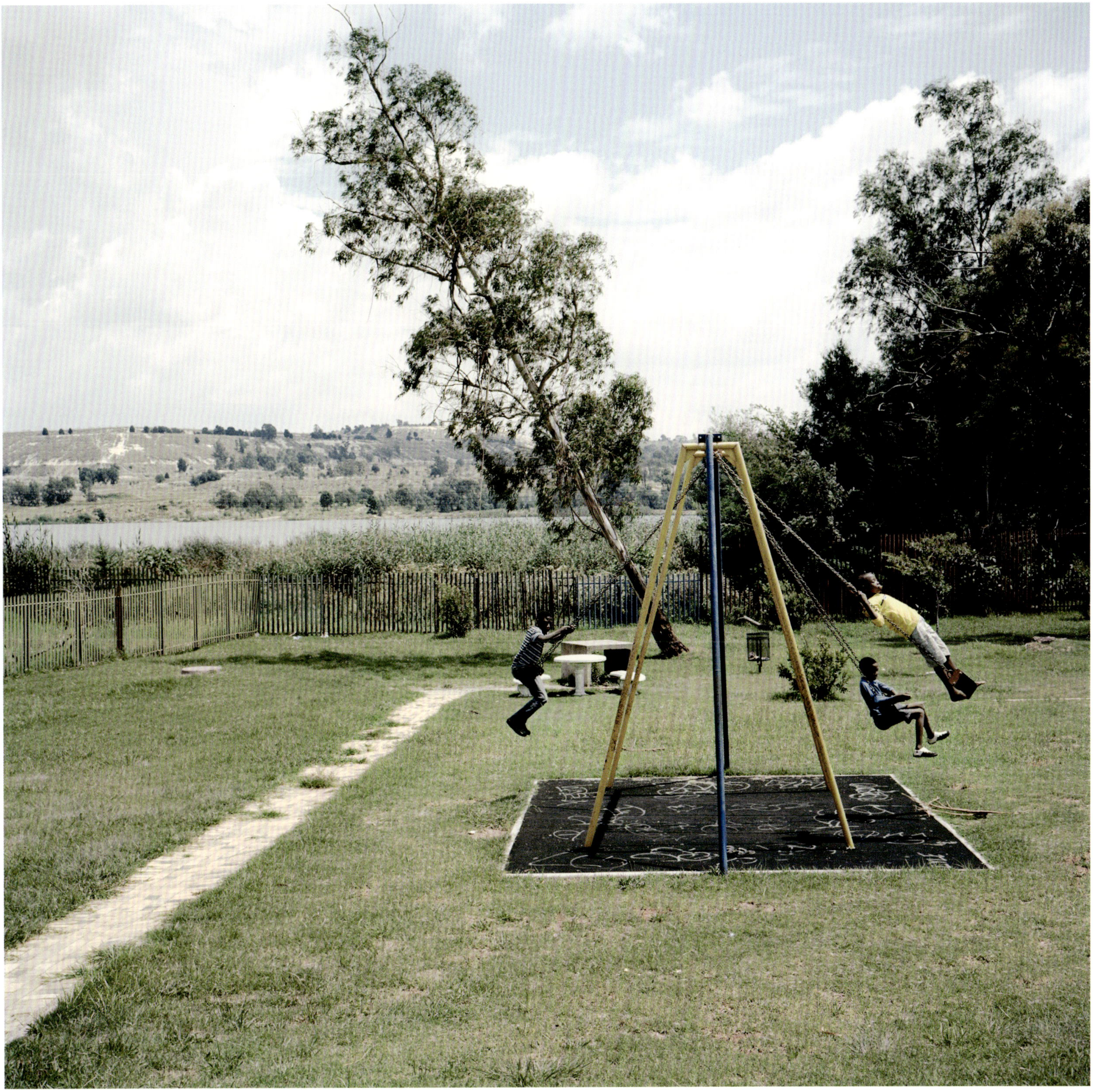

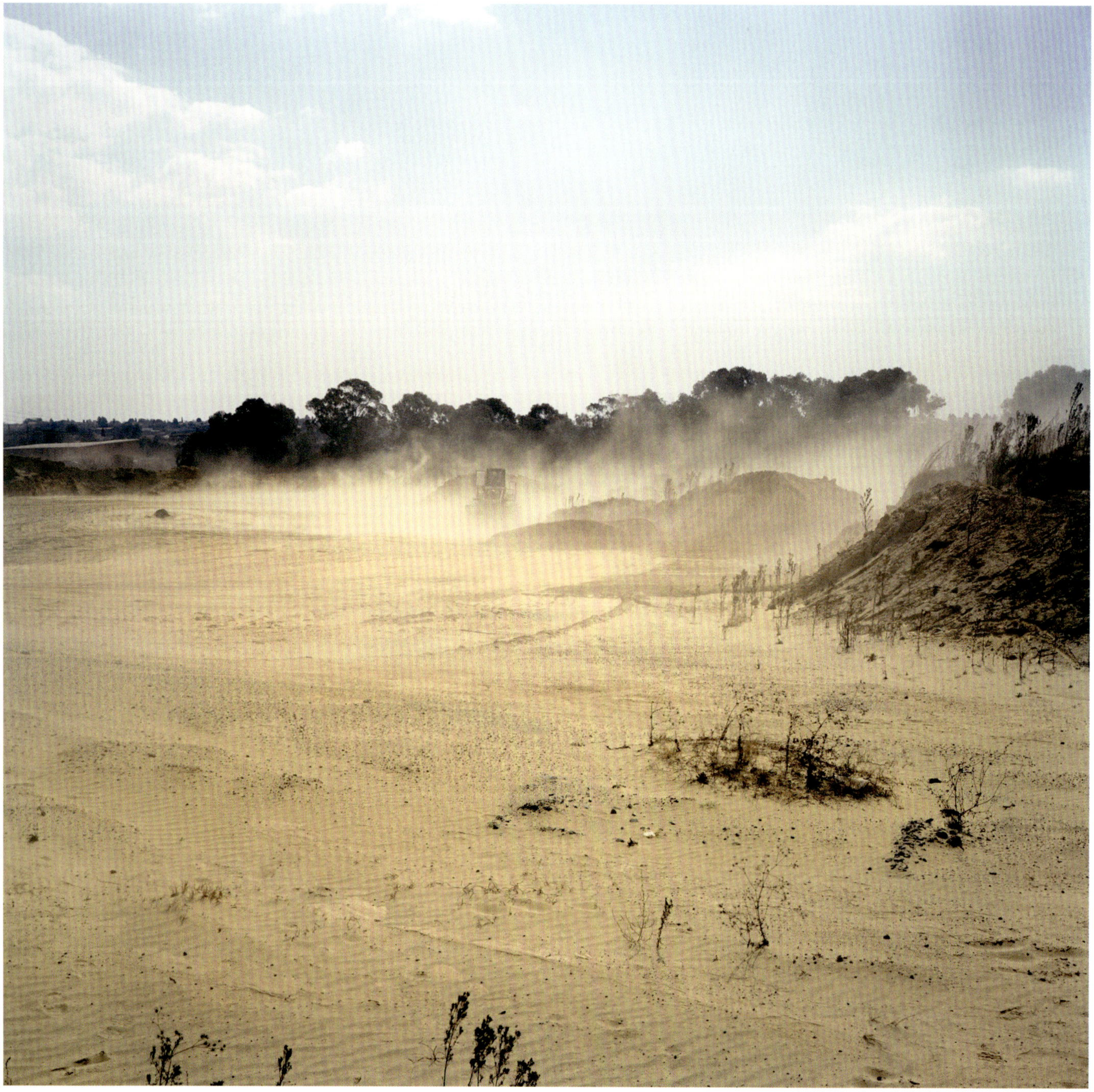

WIND-SWEPT SAND
Germiston, Johannesburg, 2010

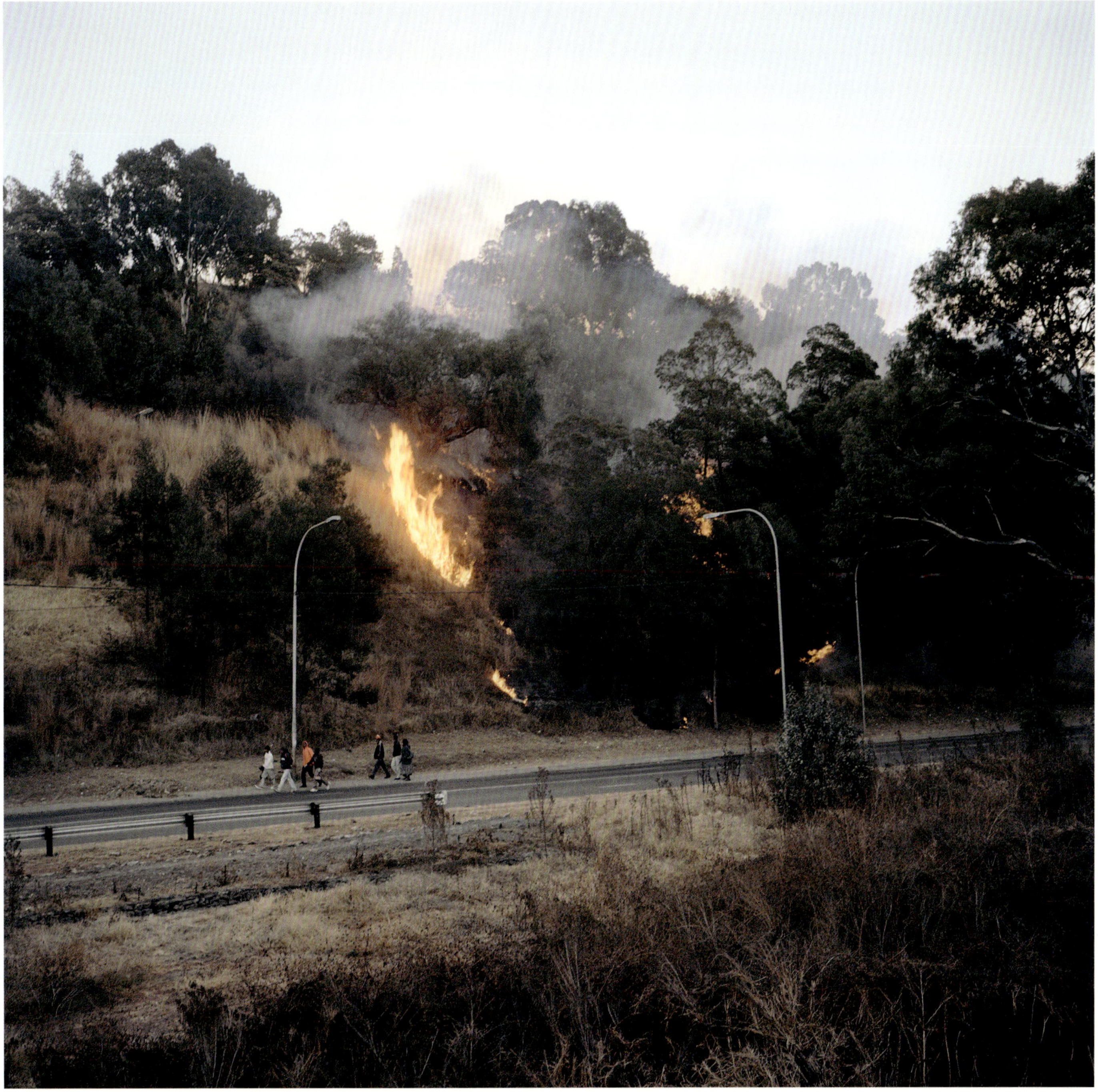

WINTER FIRES
City Deep, Johannesburg, 2010

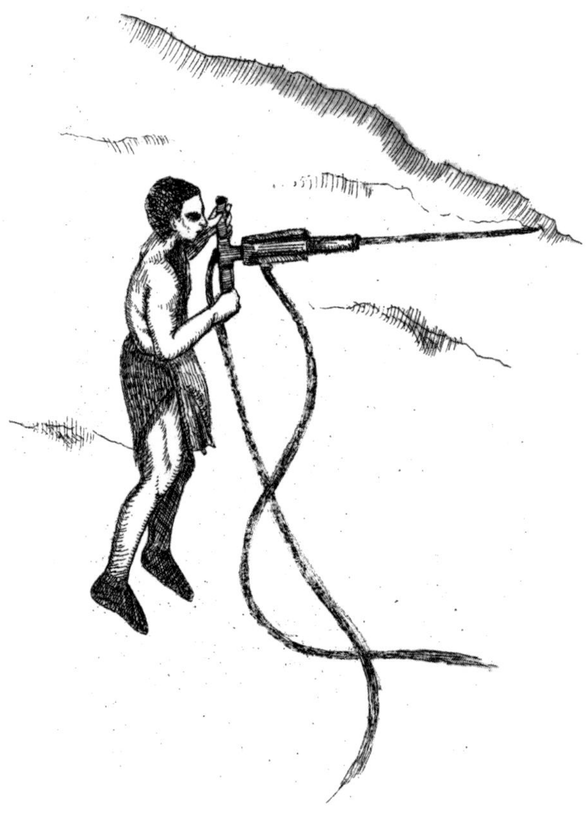

“hayikona ngena ku lo sayid pambili
ka lo skati lo bassboy zo tshela wena”

COMMUNAL SUNDAY PRAYERS
City Deep, Johannesburg, 2013

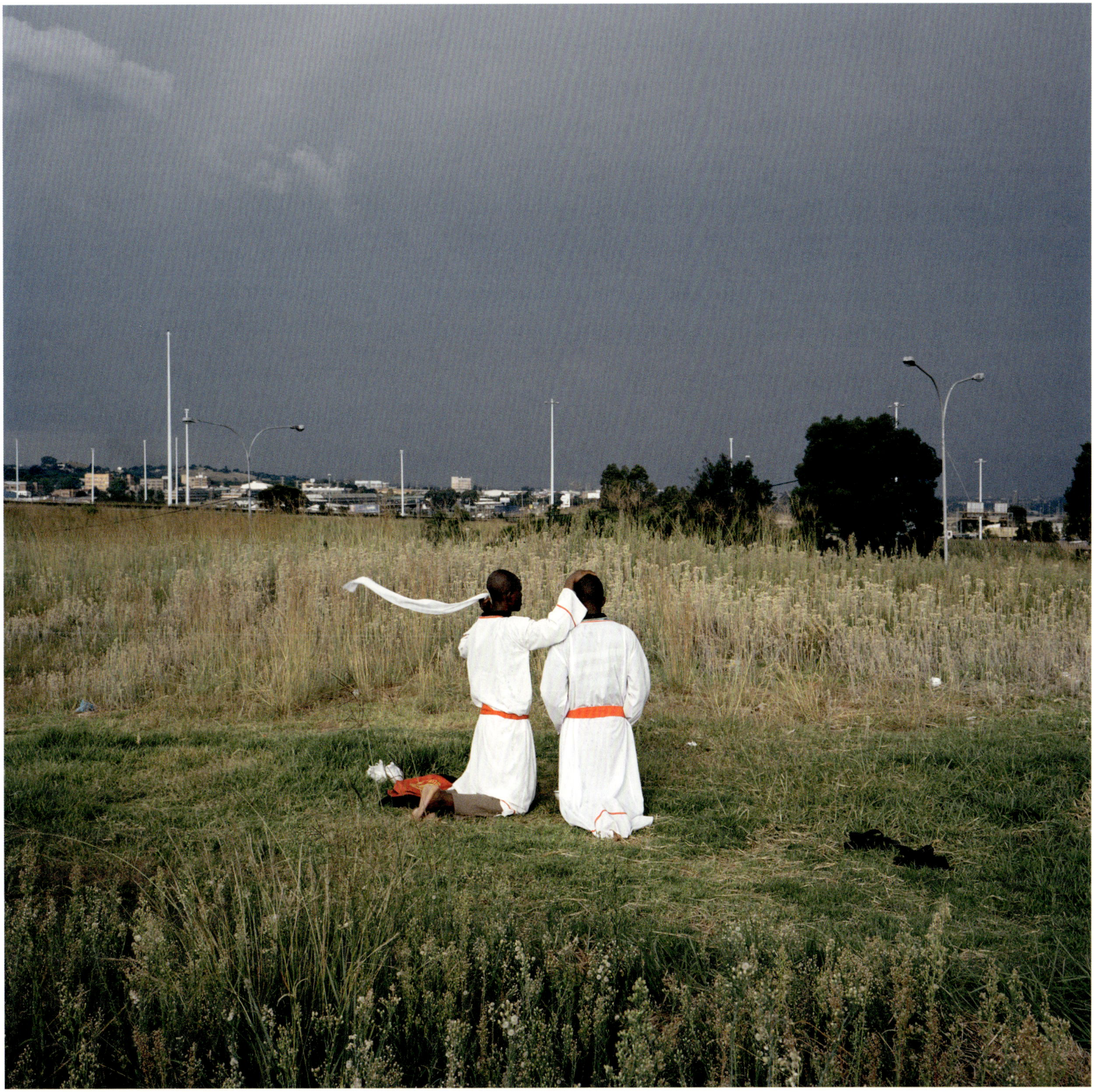

AMD
Robertville, Johannesburg, 2011

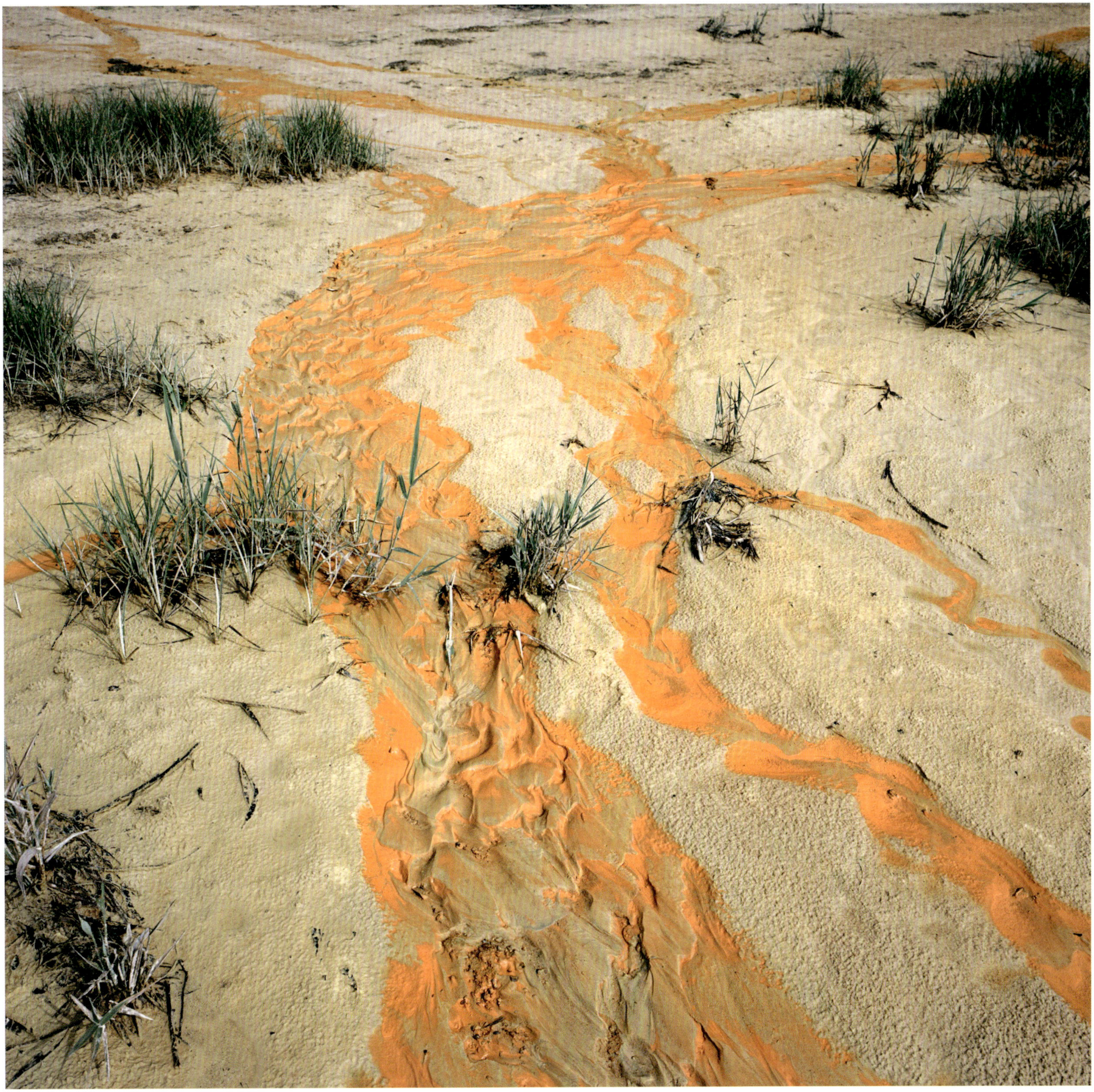

MANAGING RE-MINING
Randfontein, Johannesburg, 2013

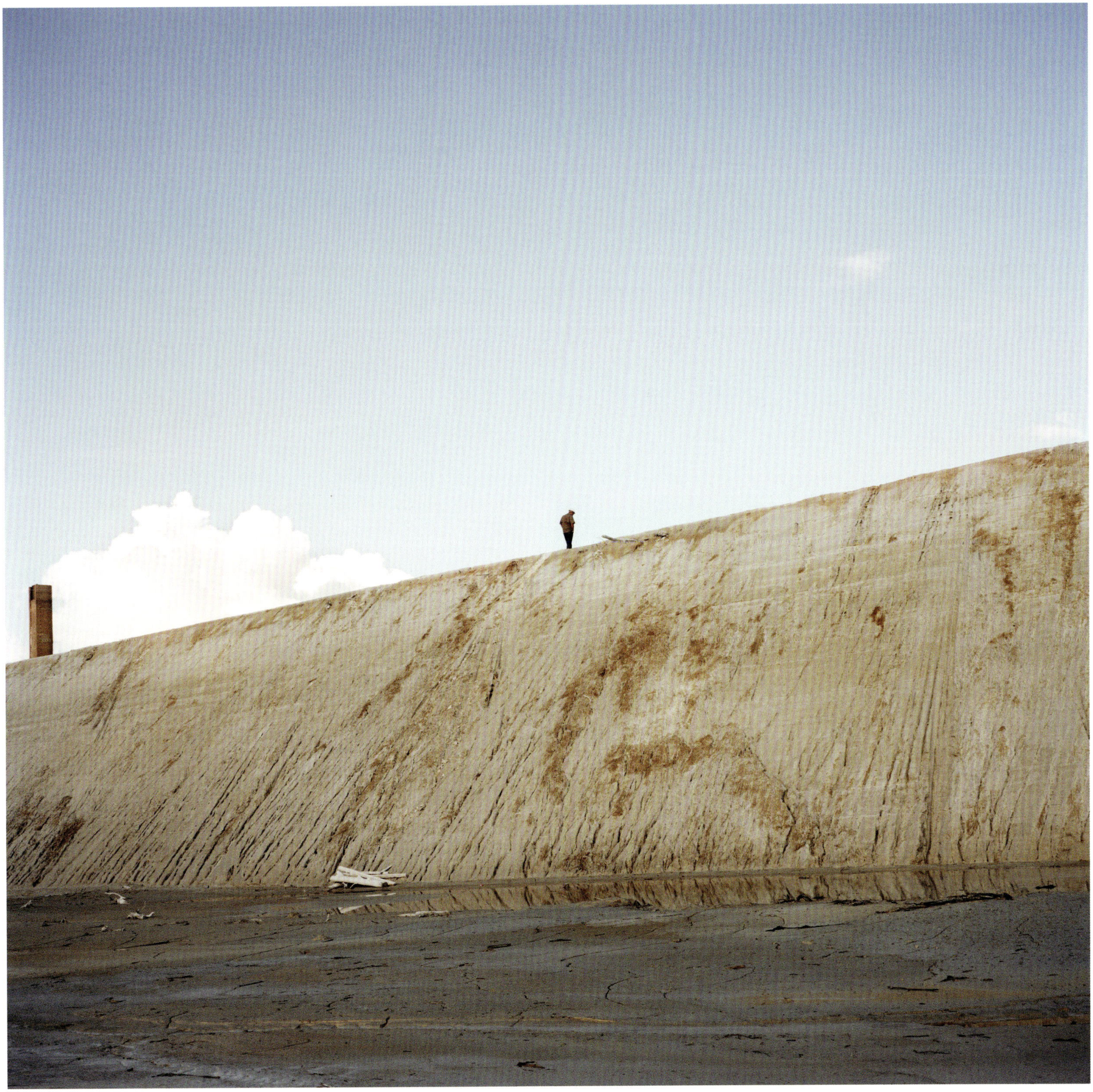

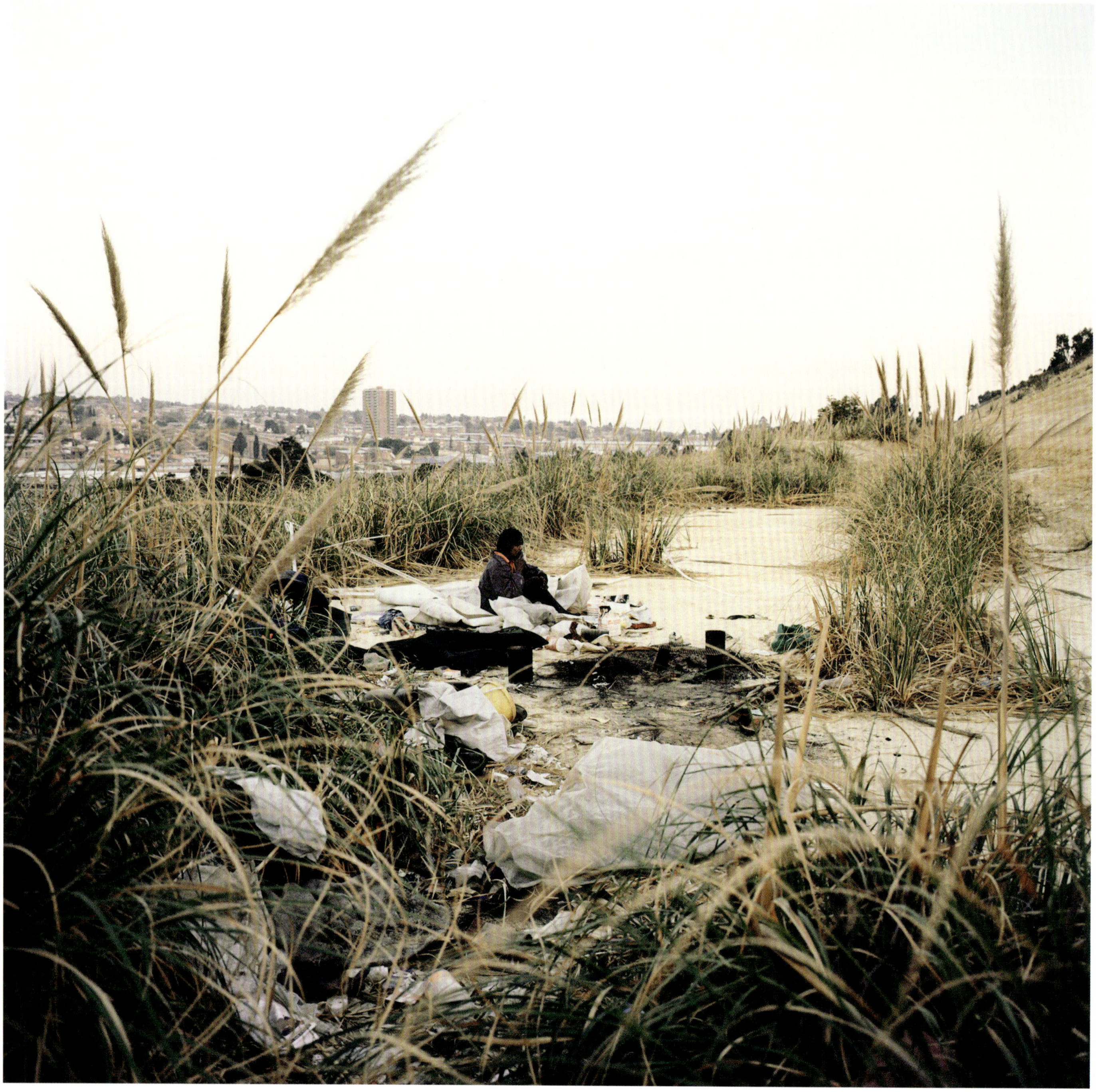

HOMELESS
West Turffontein, Johannesburg, 2013

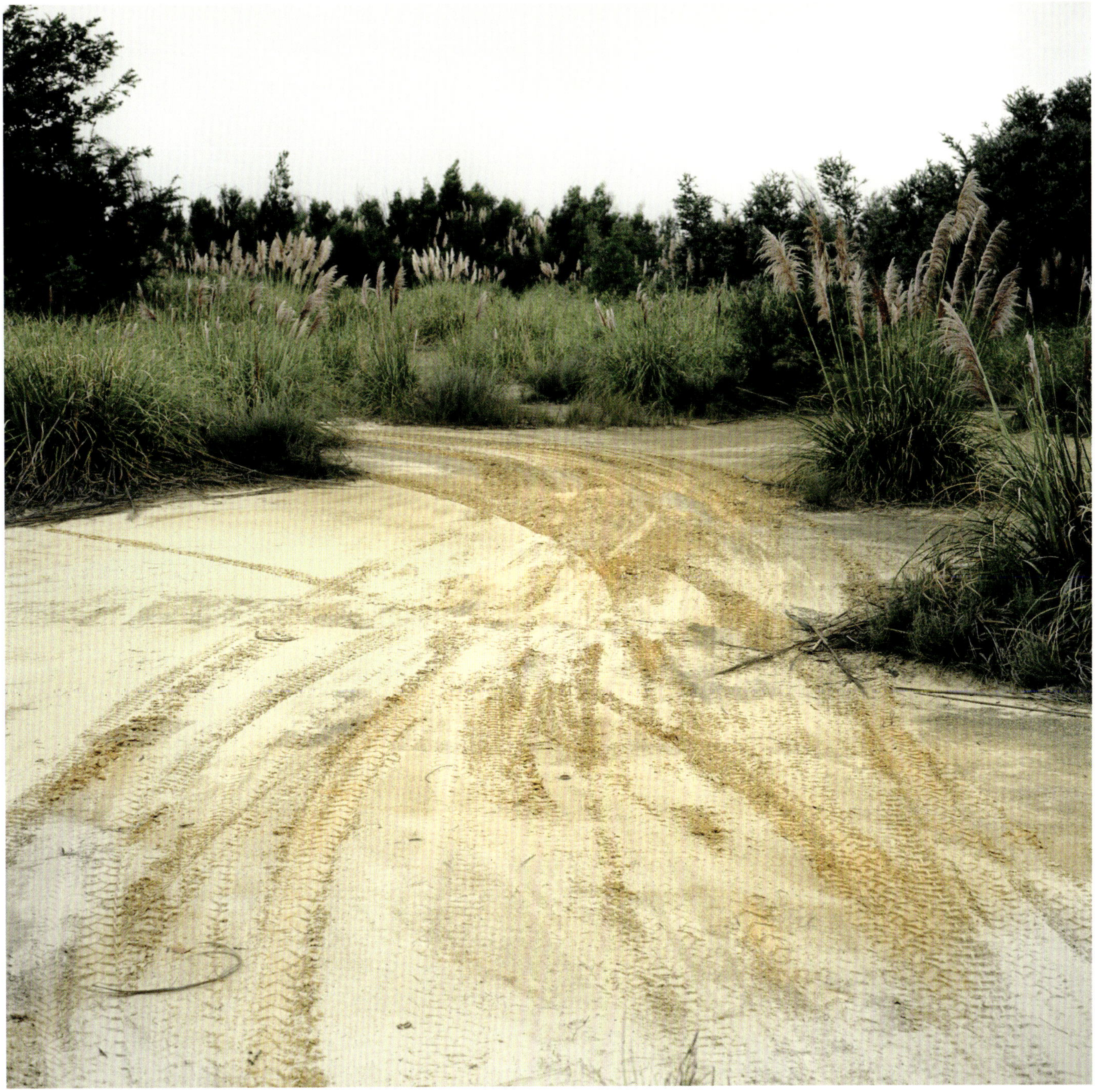

MOTORBIKE TRACKS
Chrisville, Johannesburg, 2013

COMMERCIAL RUBBISH DUMP
City Deep, Johannesburg, 2010

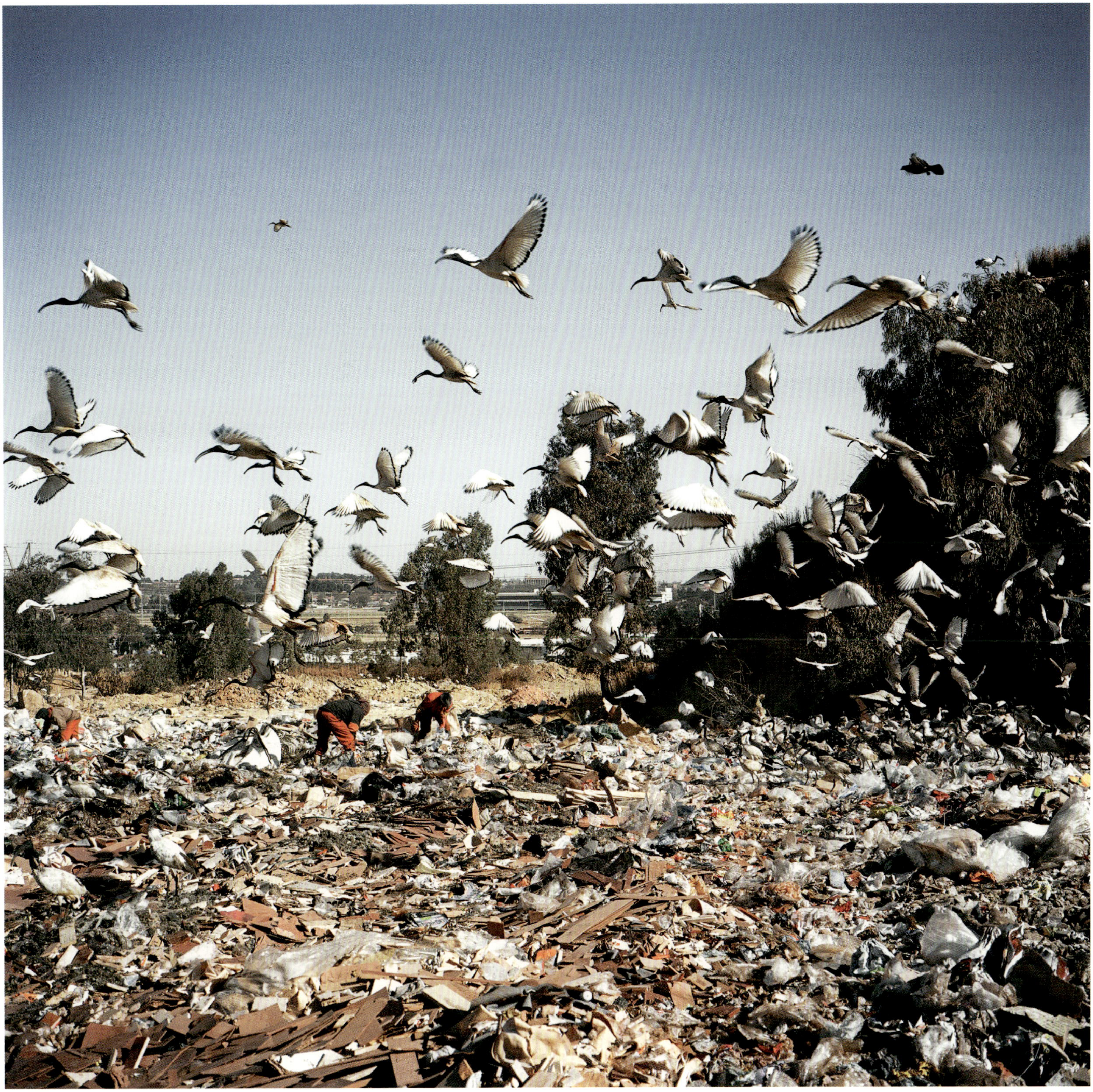

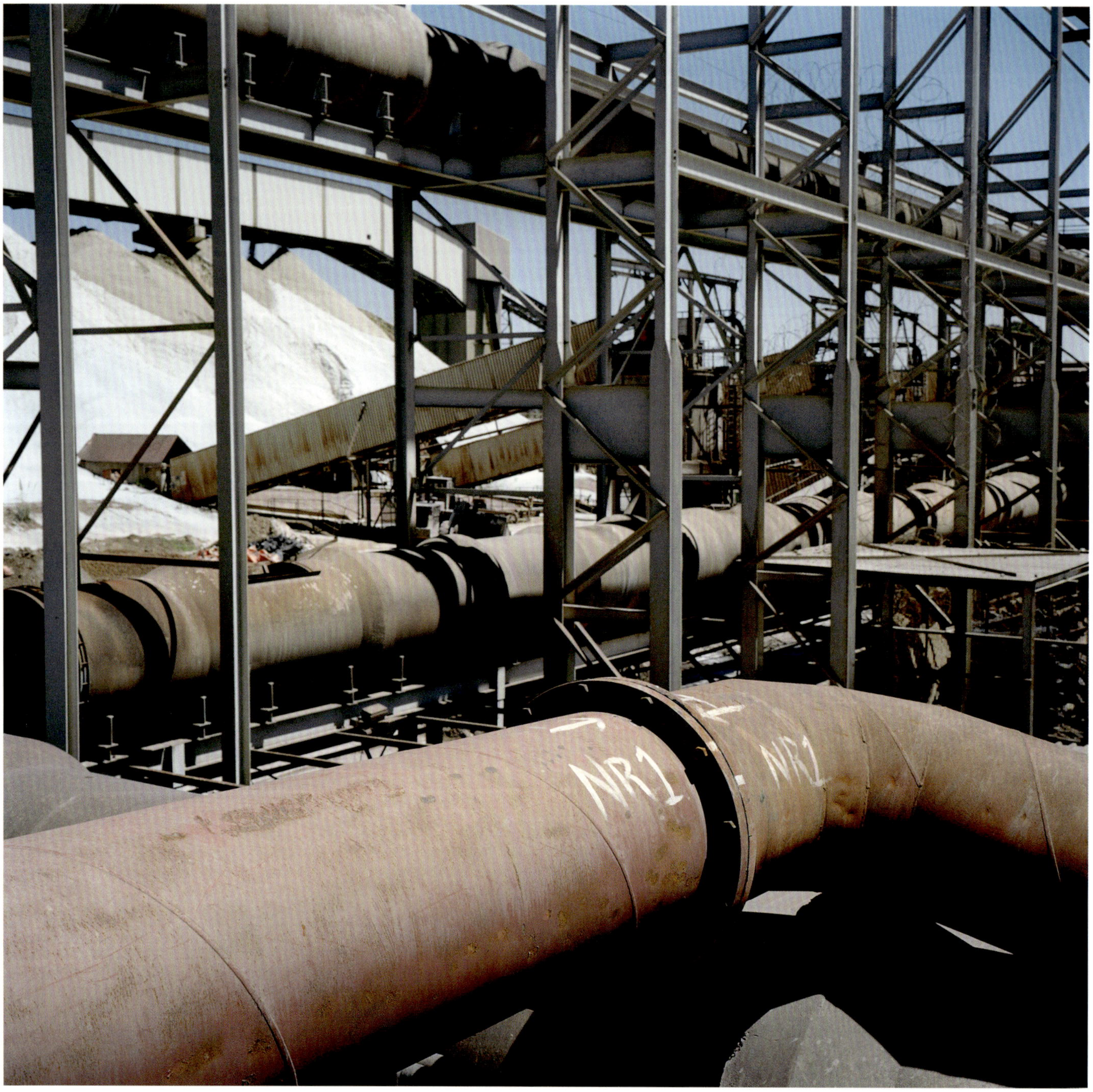

INDUSTRIAL EXTRACTION OF GOLD
Krugersdorp, Johannesburg, 2013

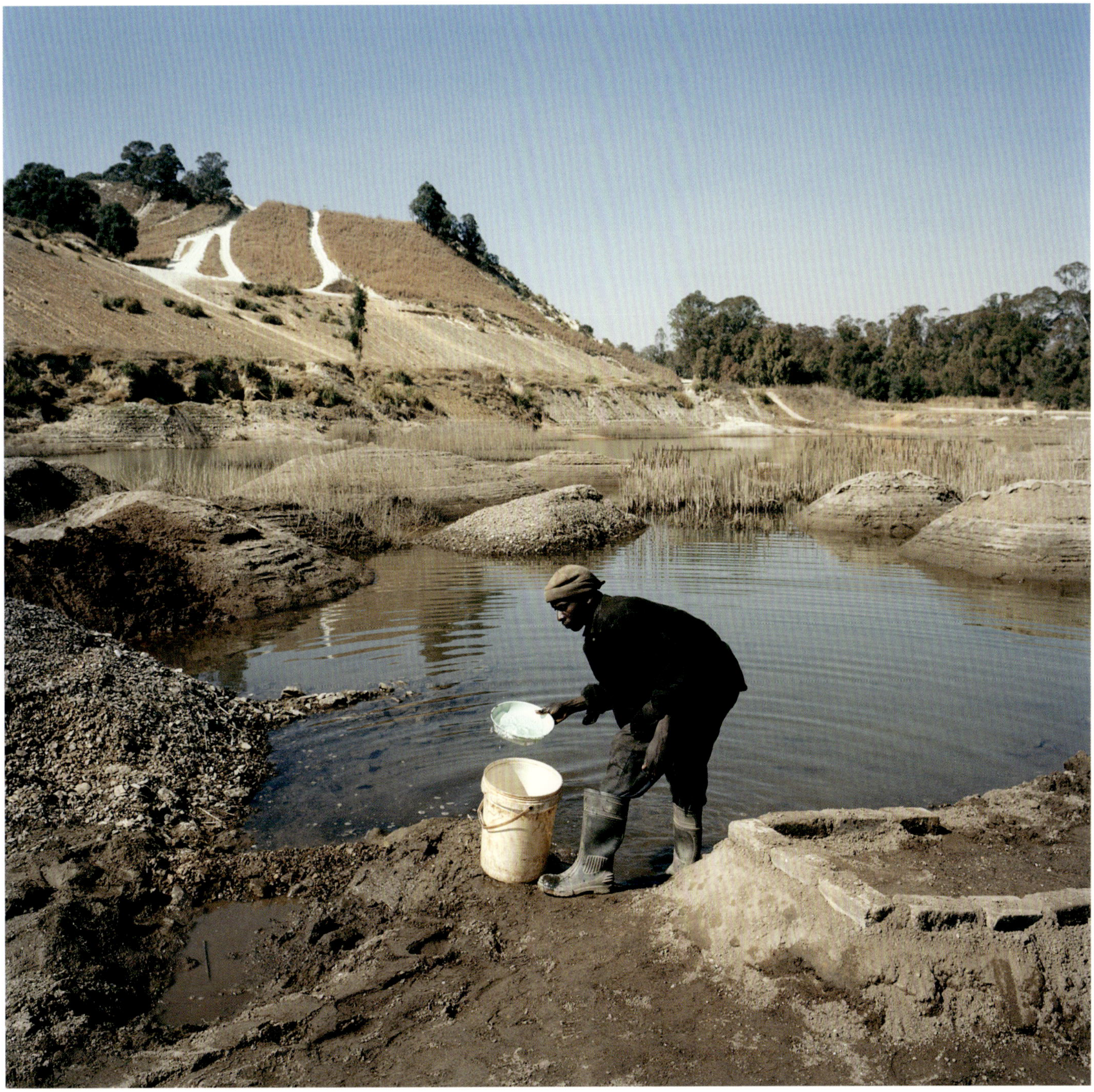

PANNING FOR GOLD
Krugersdorp, Johannesburg, 2010

WINDBREAKERS
Soweto, Johannesburg, 2013

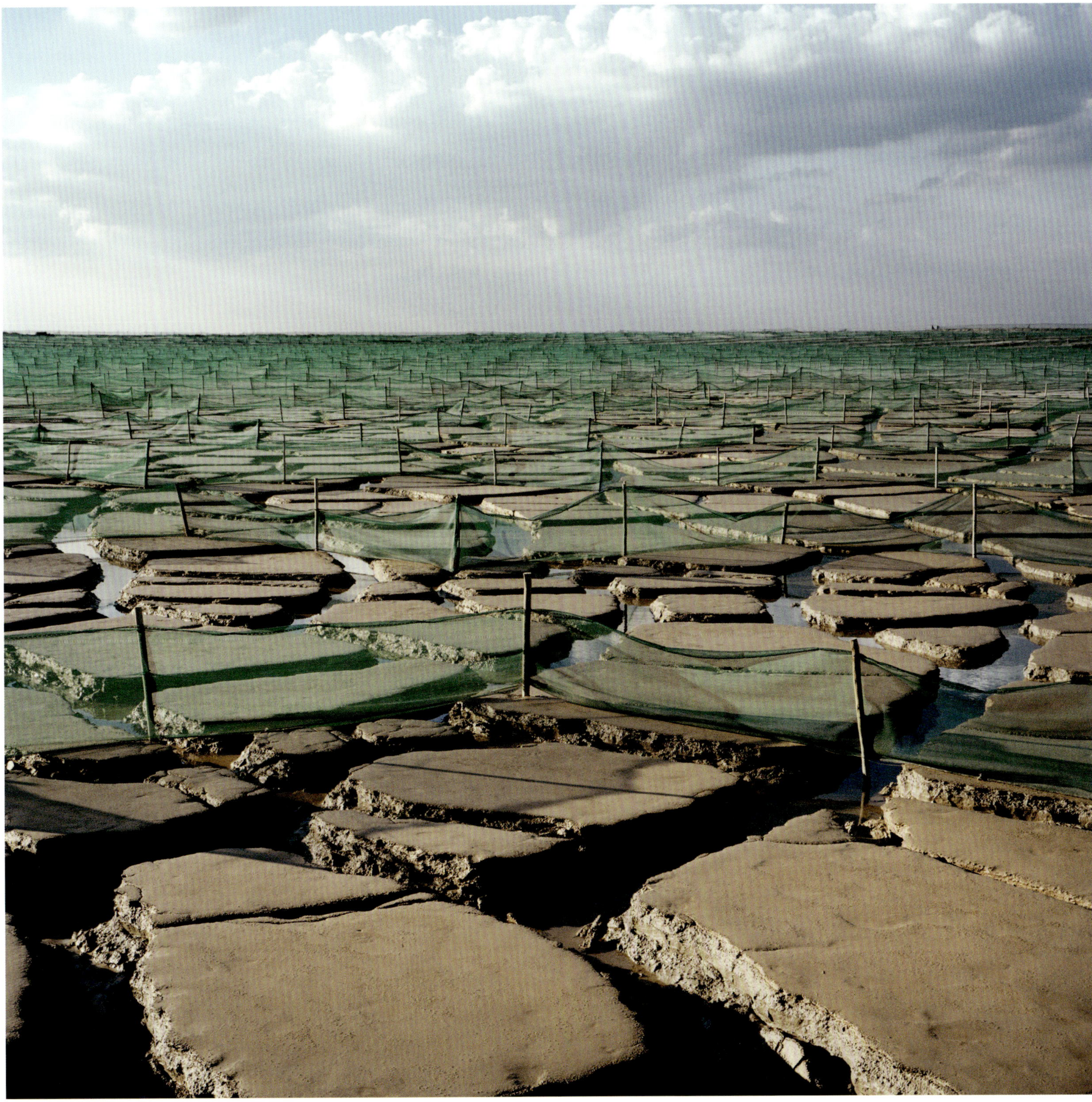

NICOLLE PRAYS BEFORE COLLECTING WATER
Meadowlands East, Johannesburg, 2012

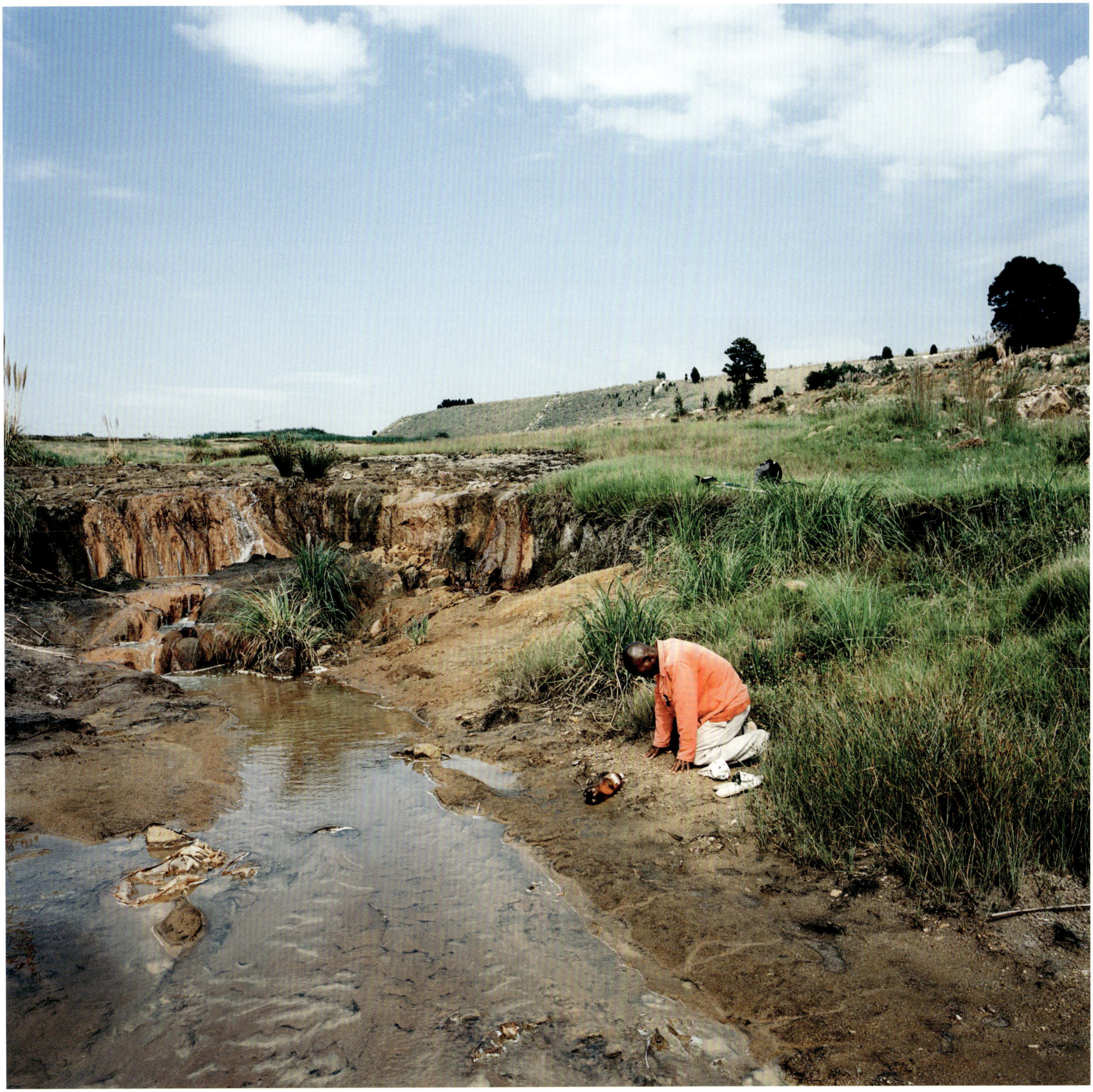

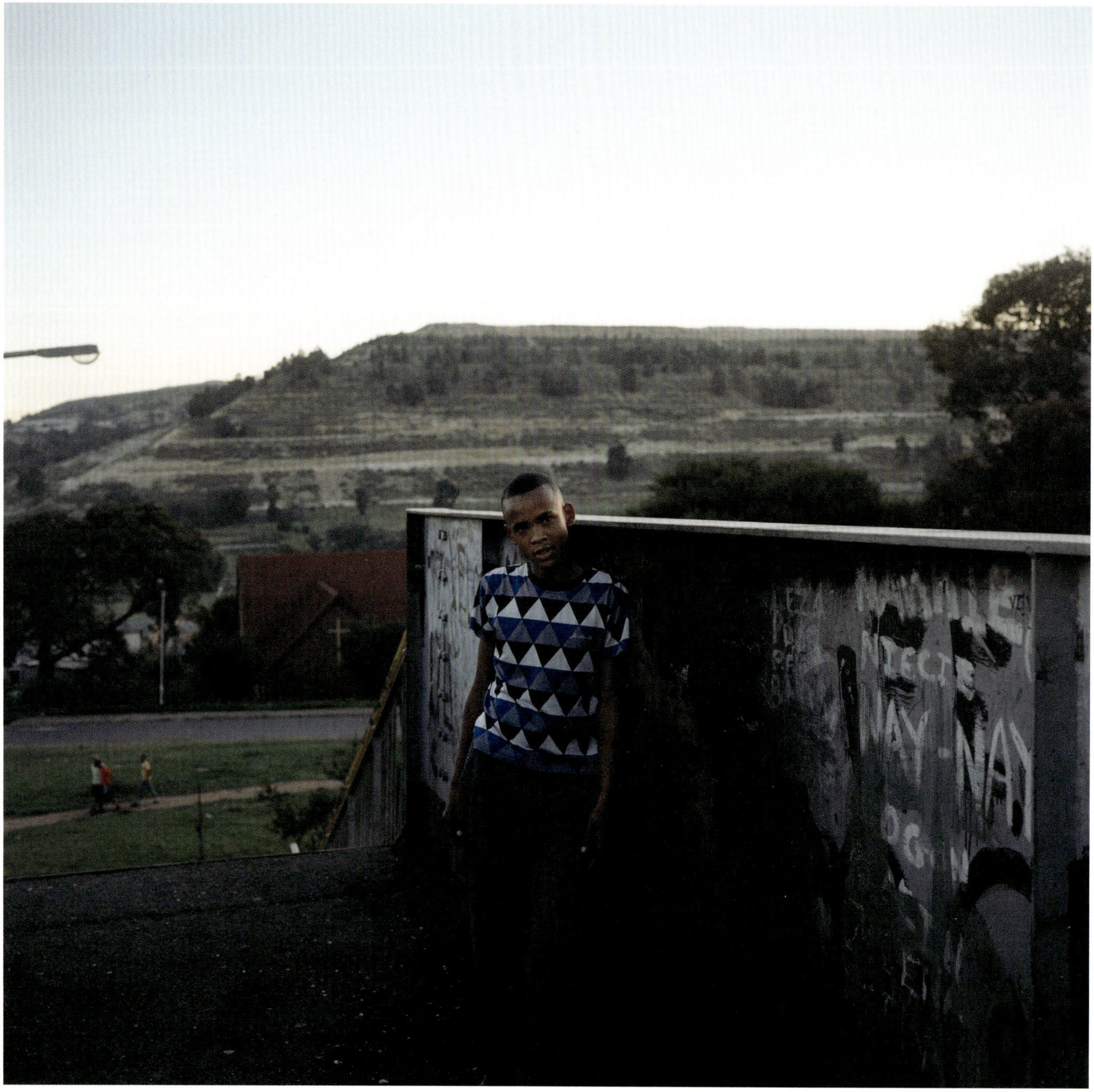

SAM
Riverlea, Johannesburg, 2013

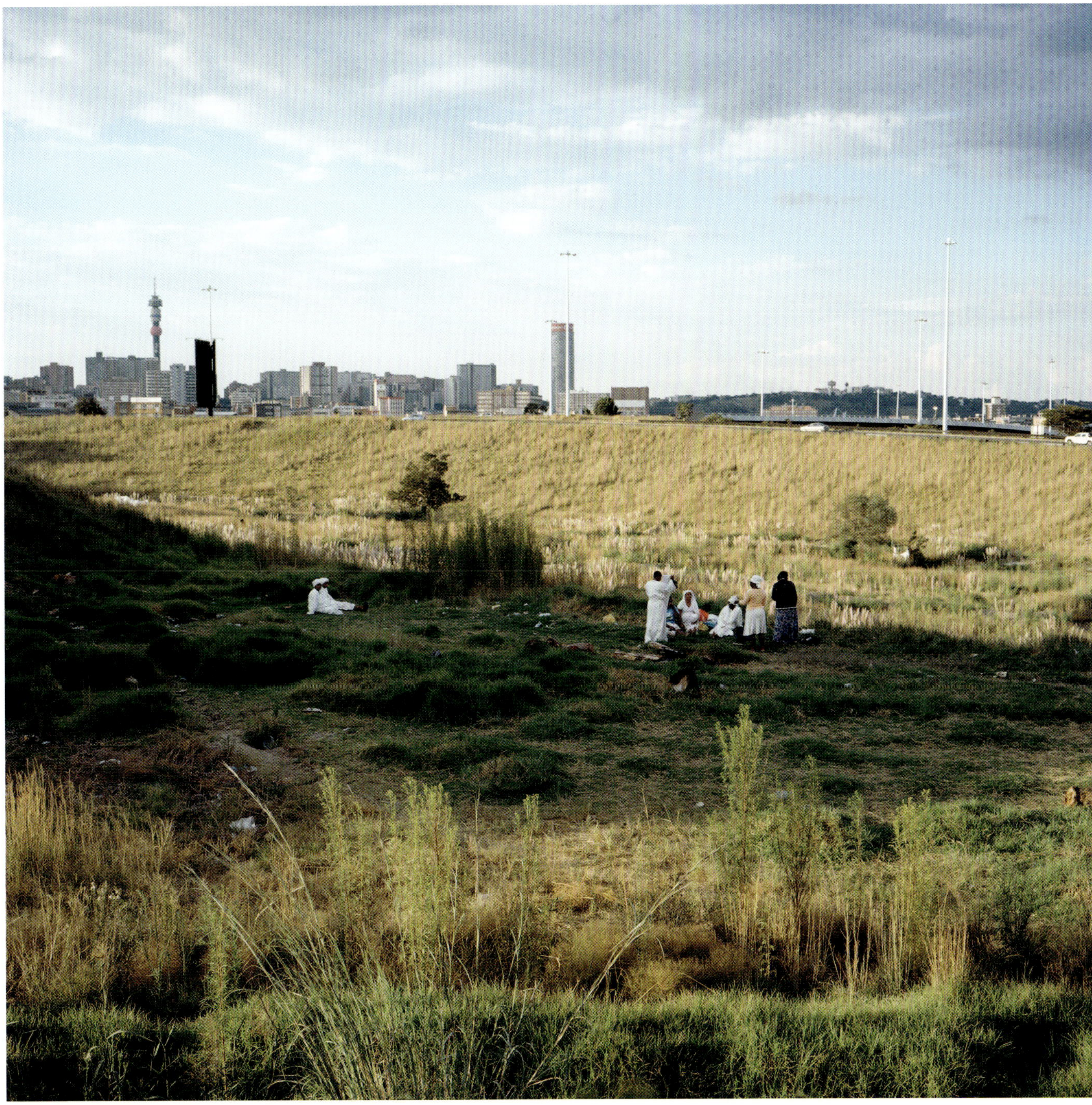

COMMUNAL SUNDAY PRAYERS
City Deep, Johannesburg, 2013

OVERLOOKING THE COMMUNITY OF DAVIDSONVILLE
Roodeport, Johannesburg, 2012

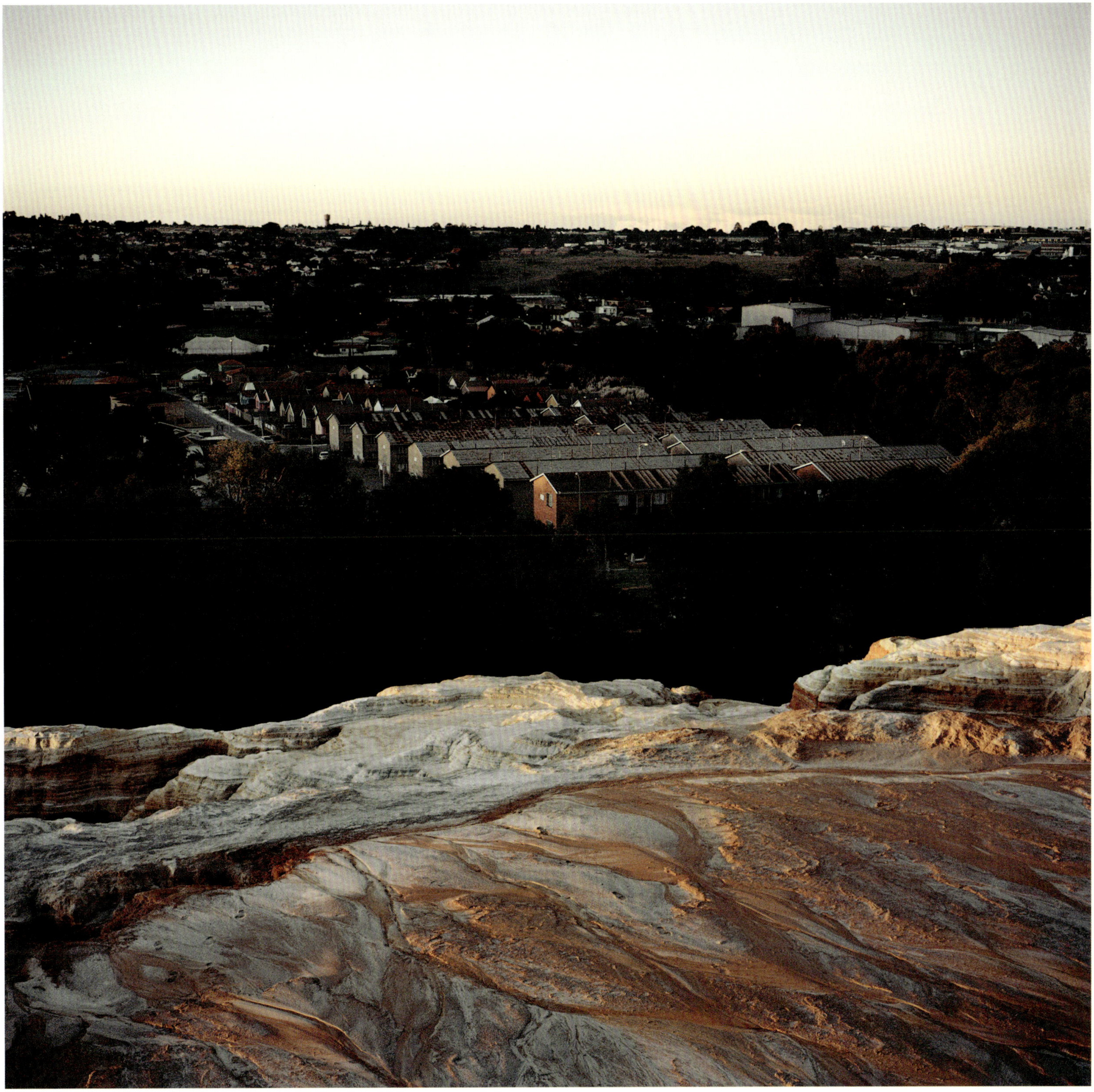

ADMI RECYCLES OFF A RUBBISH DUMP, ON A MINE DUMP
Turffontein, Johannesburg, 2010

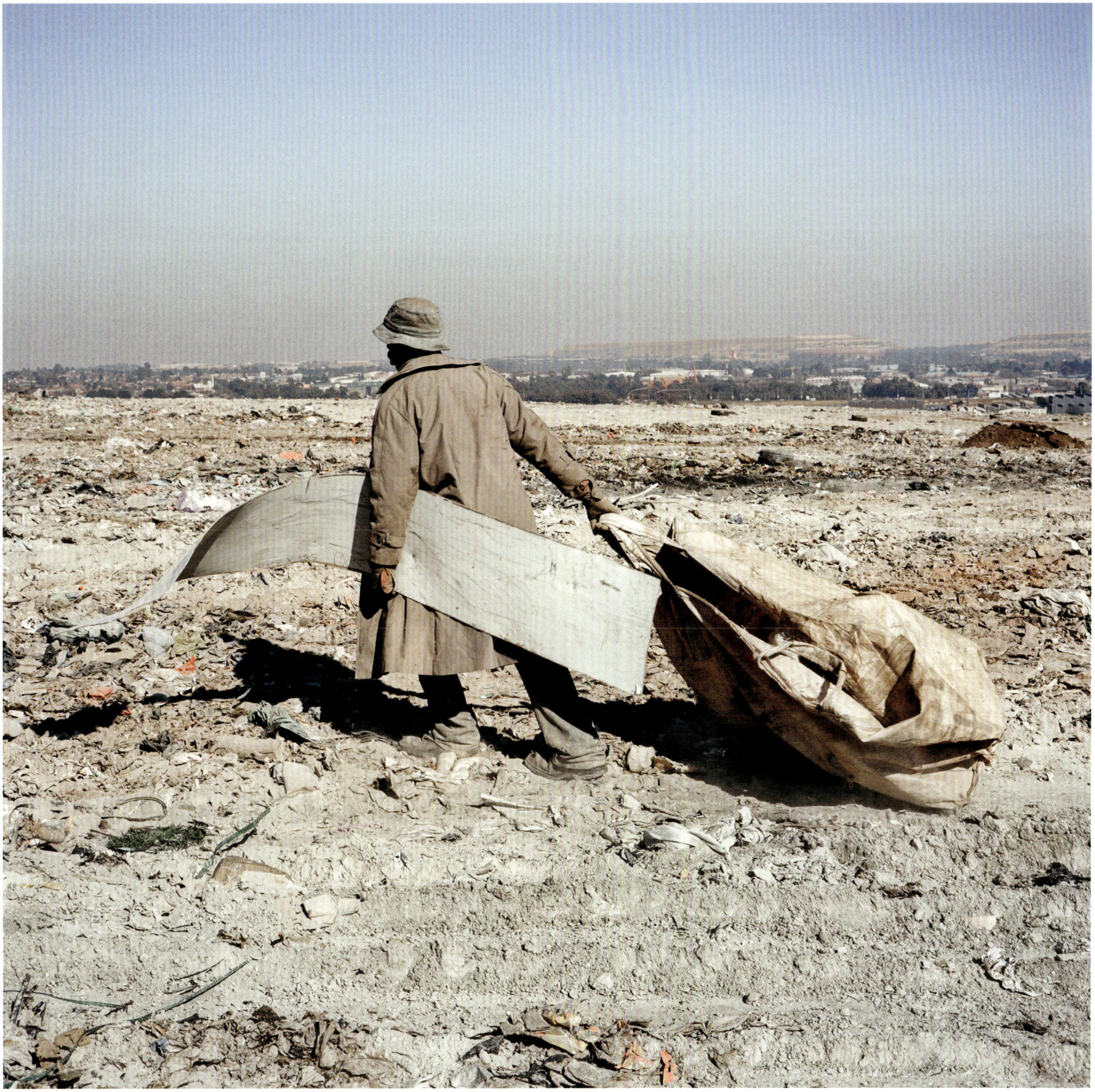

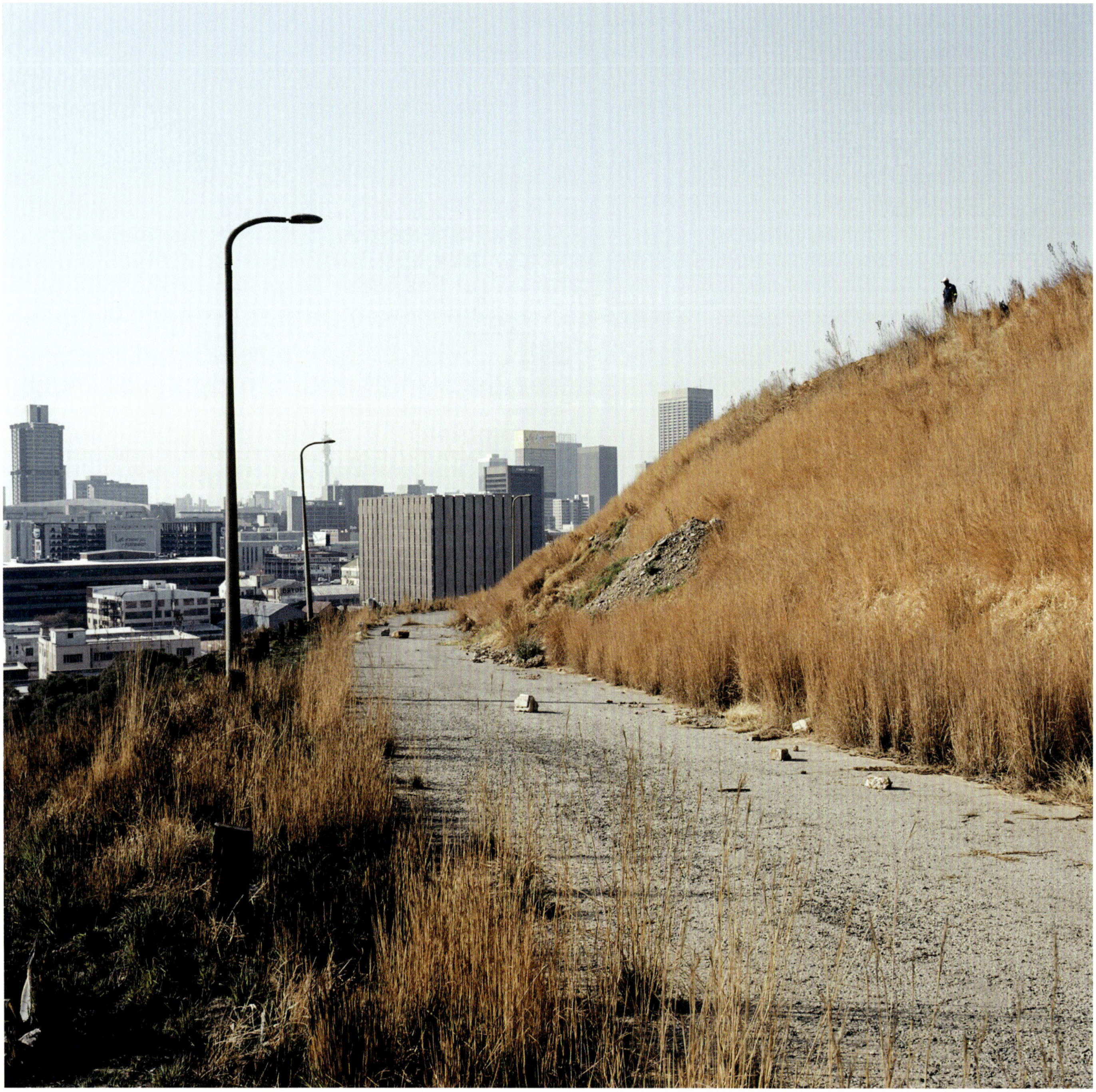

ROAD TO 'TOP STAR'
Boysens, Johannesburg, 2010

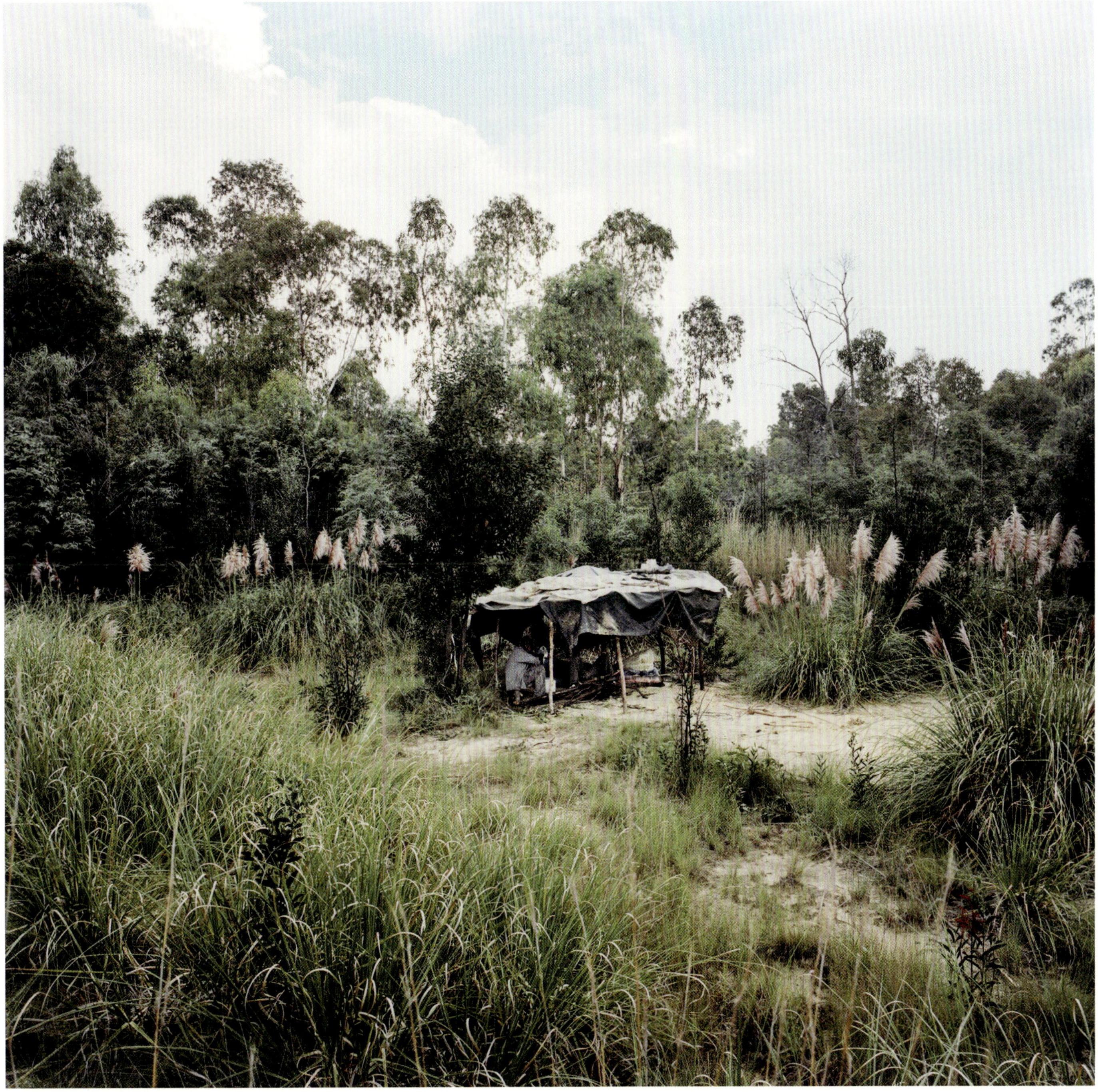

SILVESTER & TODD FASTING
West Turffontein, Johannesburg, 2012

CHILDREN AT HOME
Delmore, Johannesburg, 2012

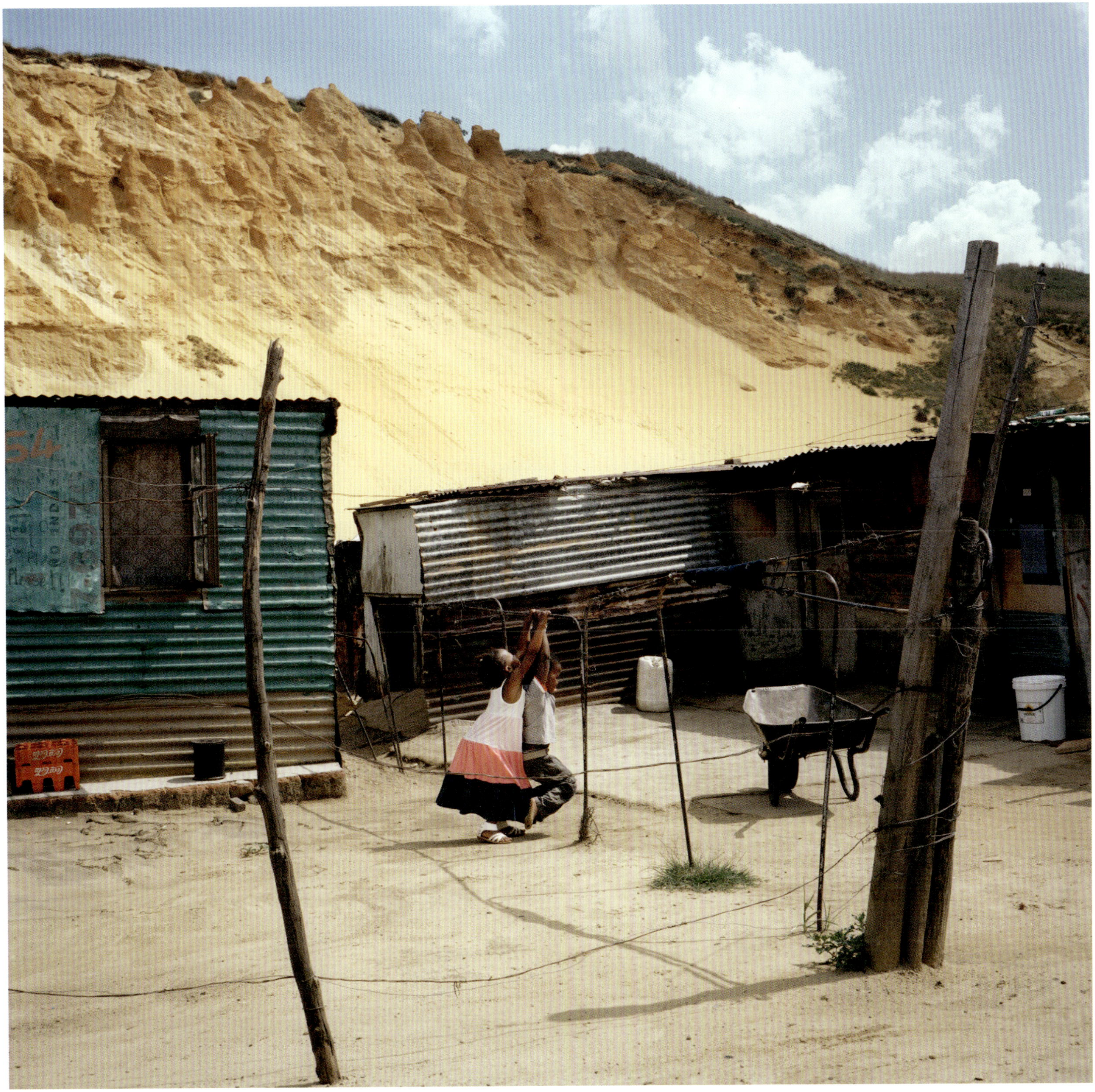
54

“mina hayiyazi kanjani yena wile
pantsi ku lo troko. yenamadala
muntu ga lo malayisha”

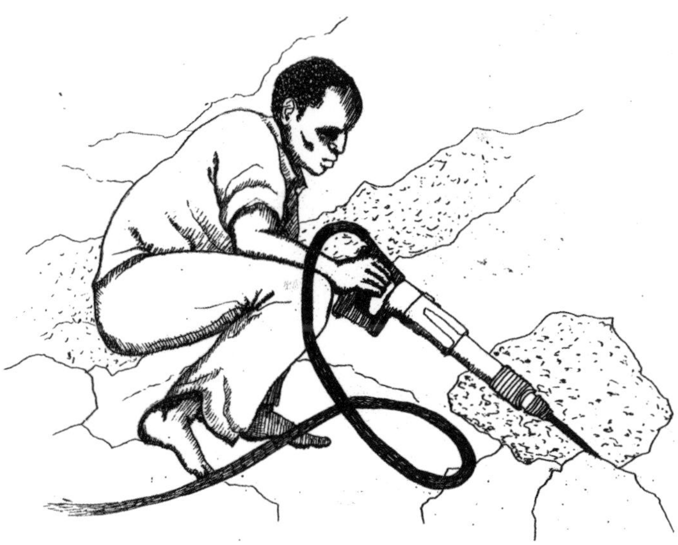

'JERUSALEM' SETTLEMENT
Delmore, Johannesburg, 2012

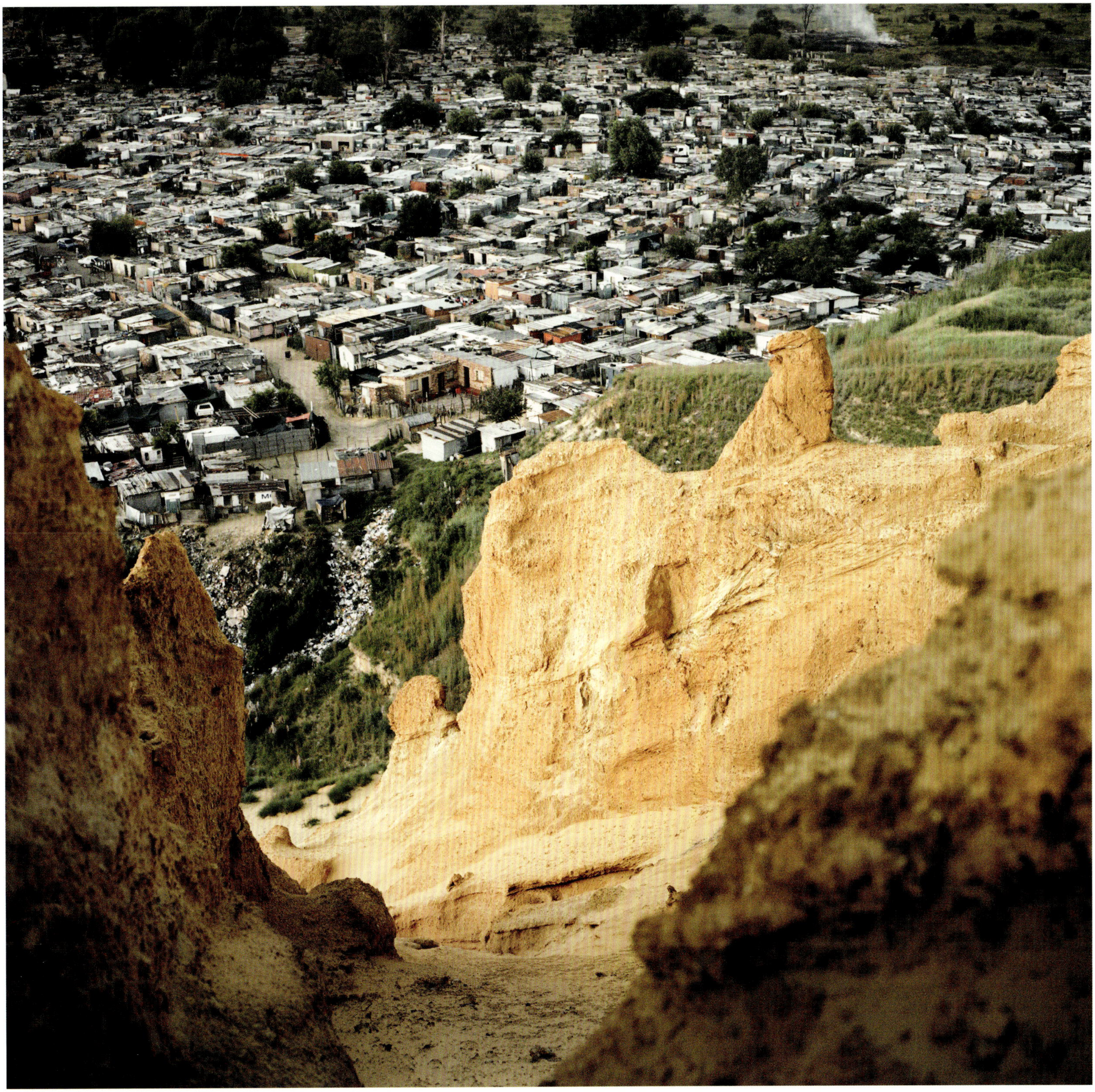

MINE DUMP SAND
Roodeport, Johannesburg, 2012

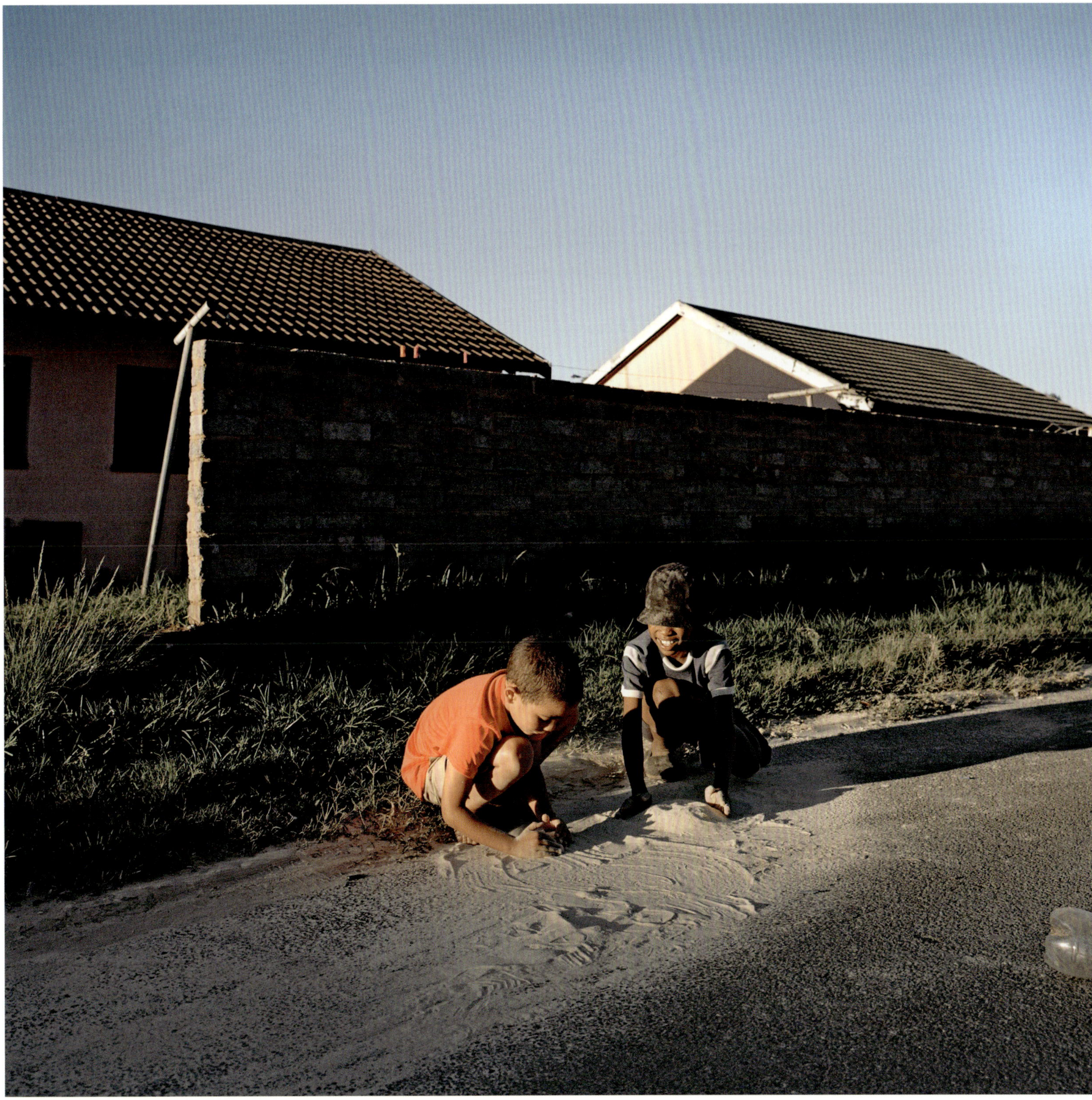

NEUTRALISED AMD
Randfontein, Johannesburg, 2012

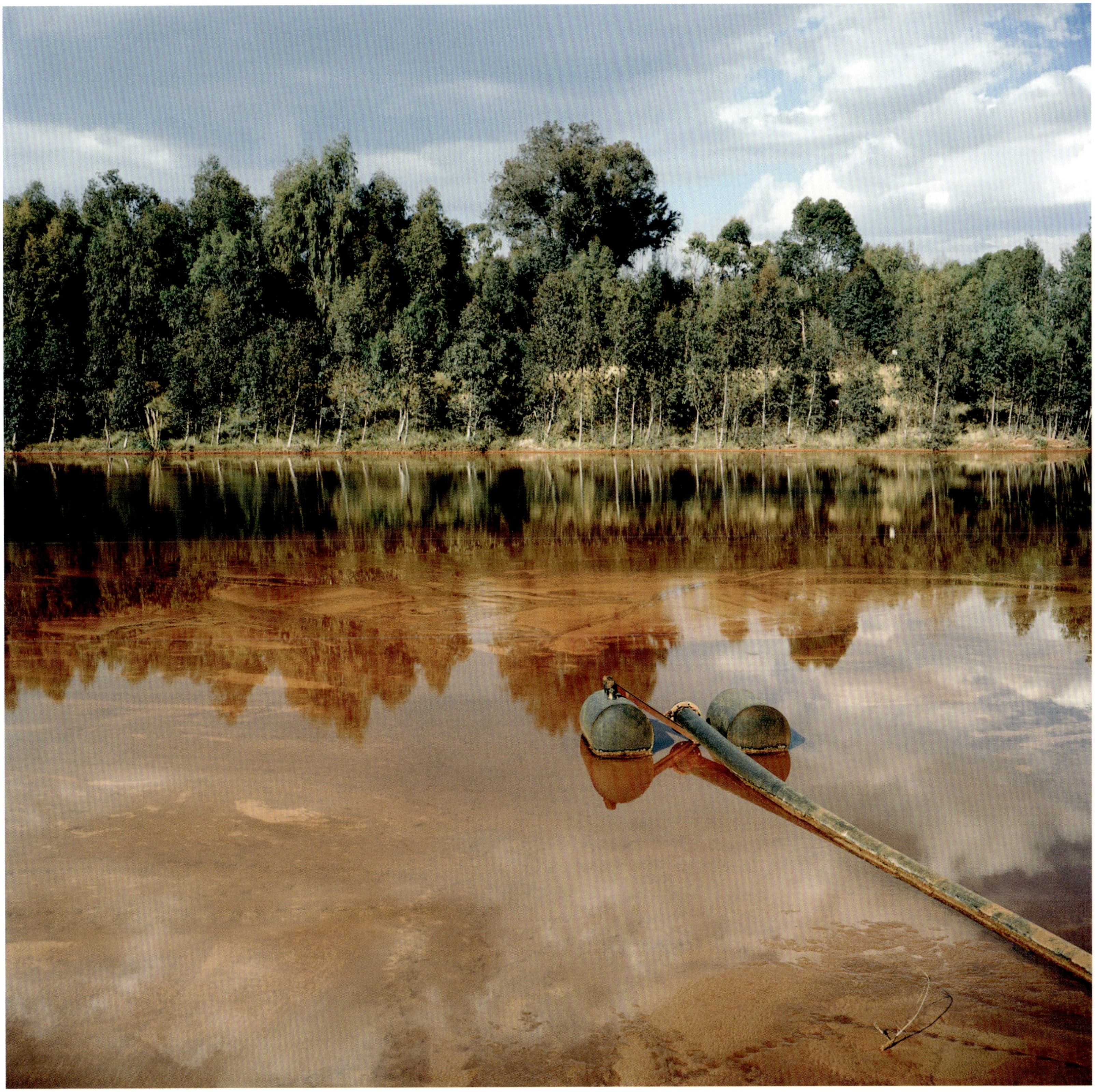

PRESSURISED WATER
Krugersdorp, Johannesburg, 2013

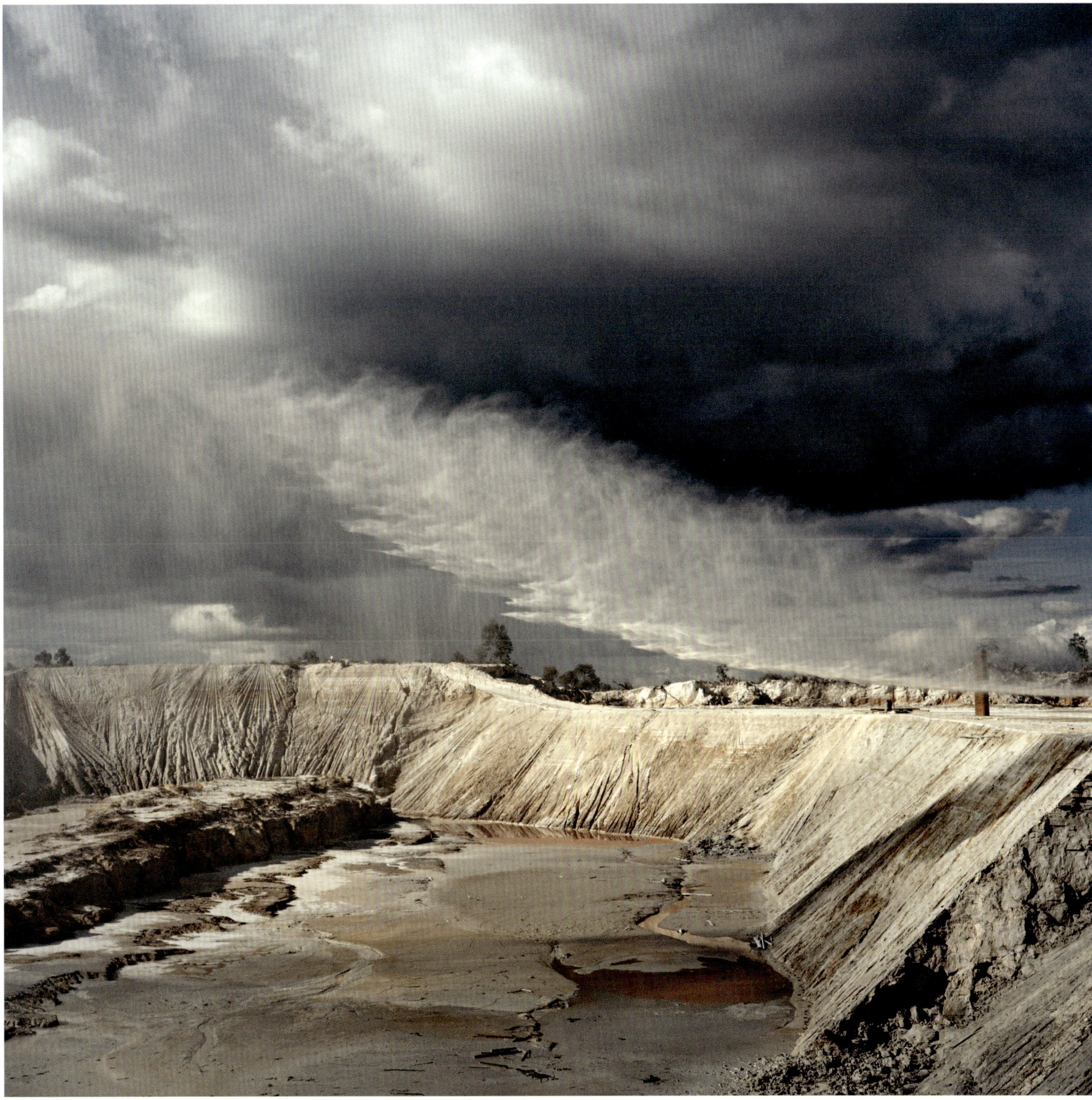

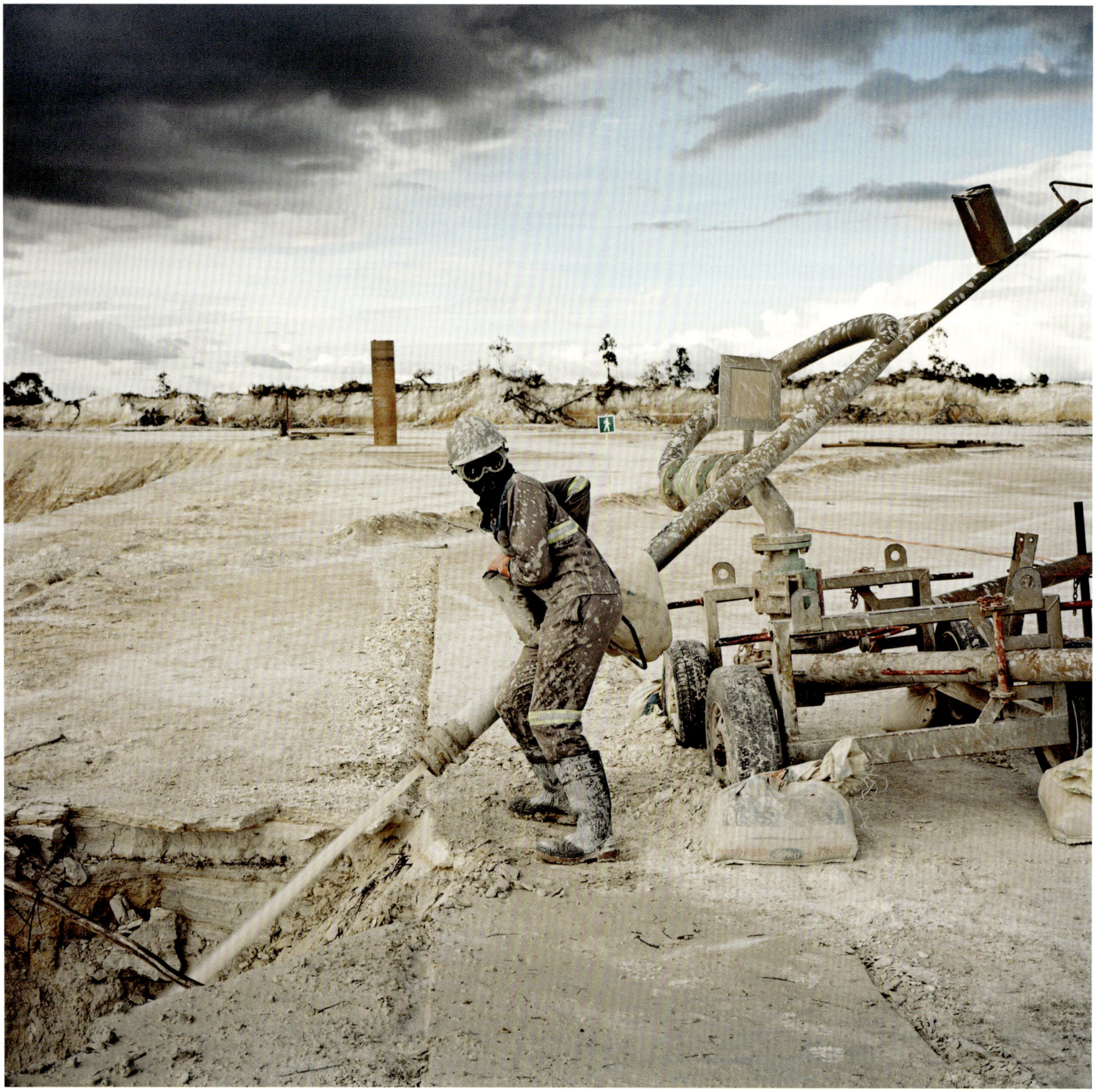

BREAKING DOWN THE DUMP
Krugersdorp, Johannesburg, 2013

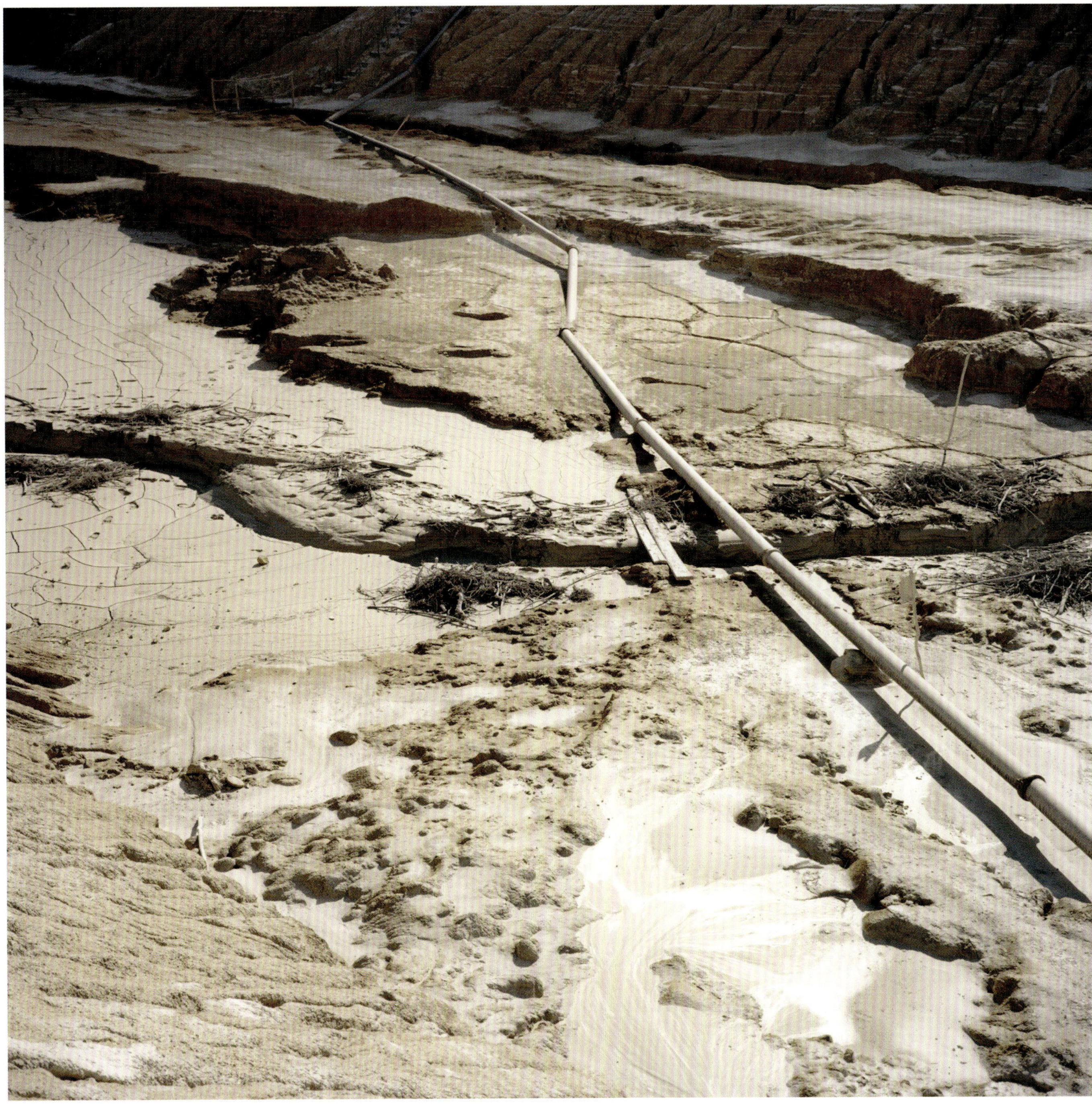

WATER PIPES
Krugersdorp, Johannesburg, 2013

A BREAK DURING PANNING
Roodepoort, Johannesburg, 2012

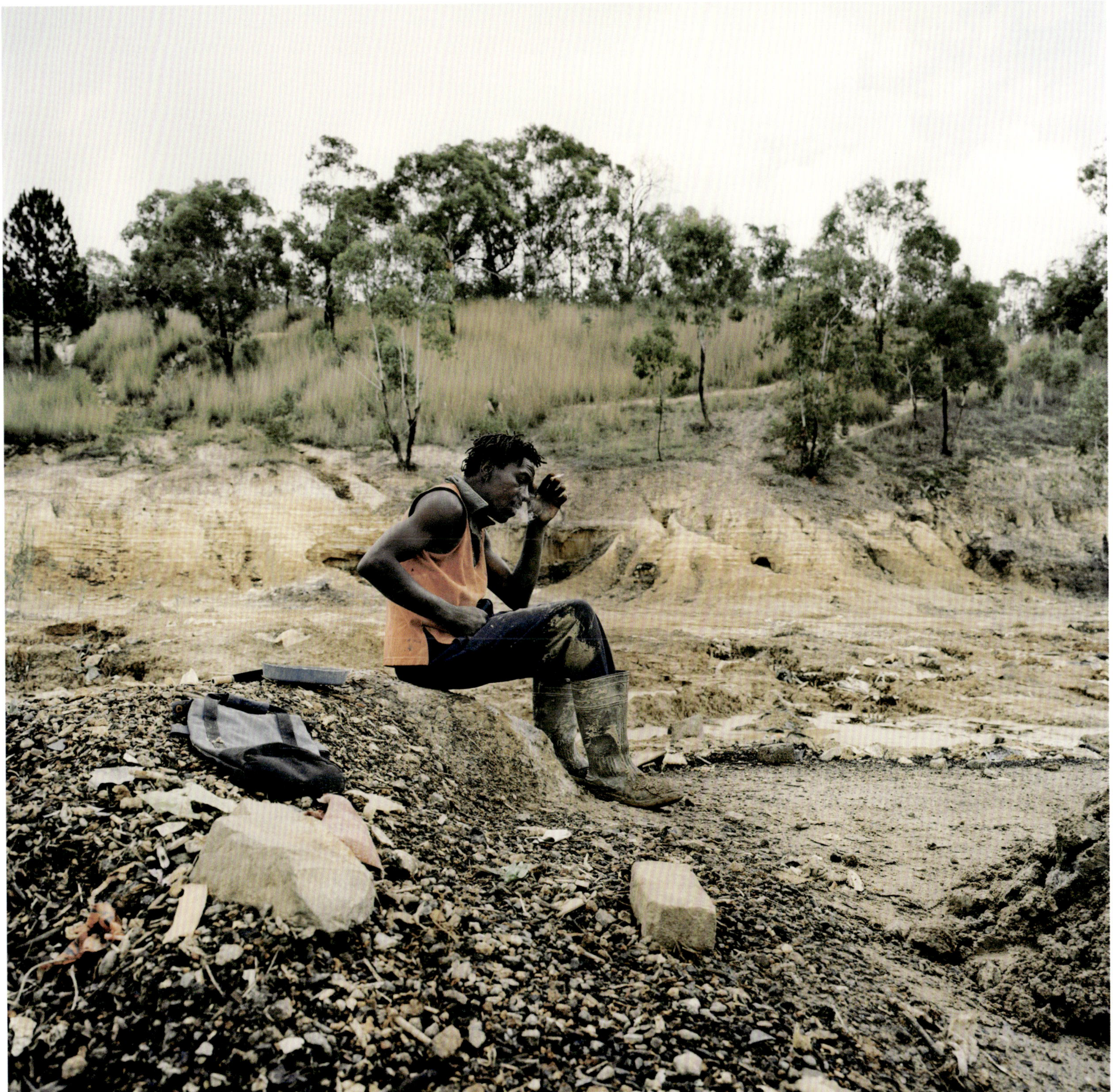

RE-MINING 'DUMP 20'
Randfontein, Johannesburg, 2012

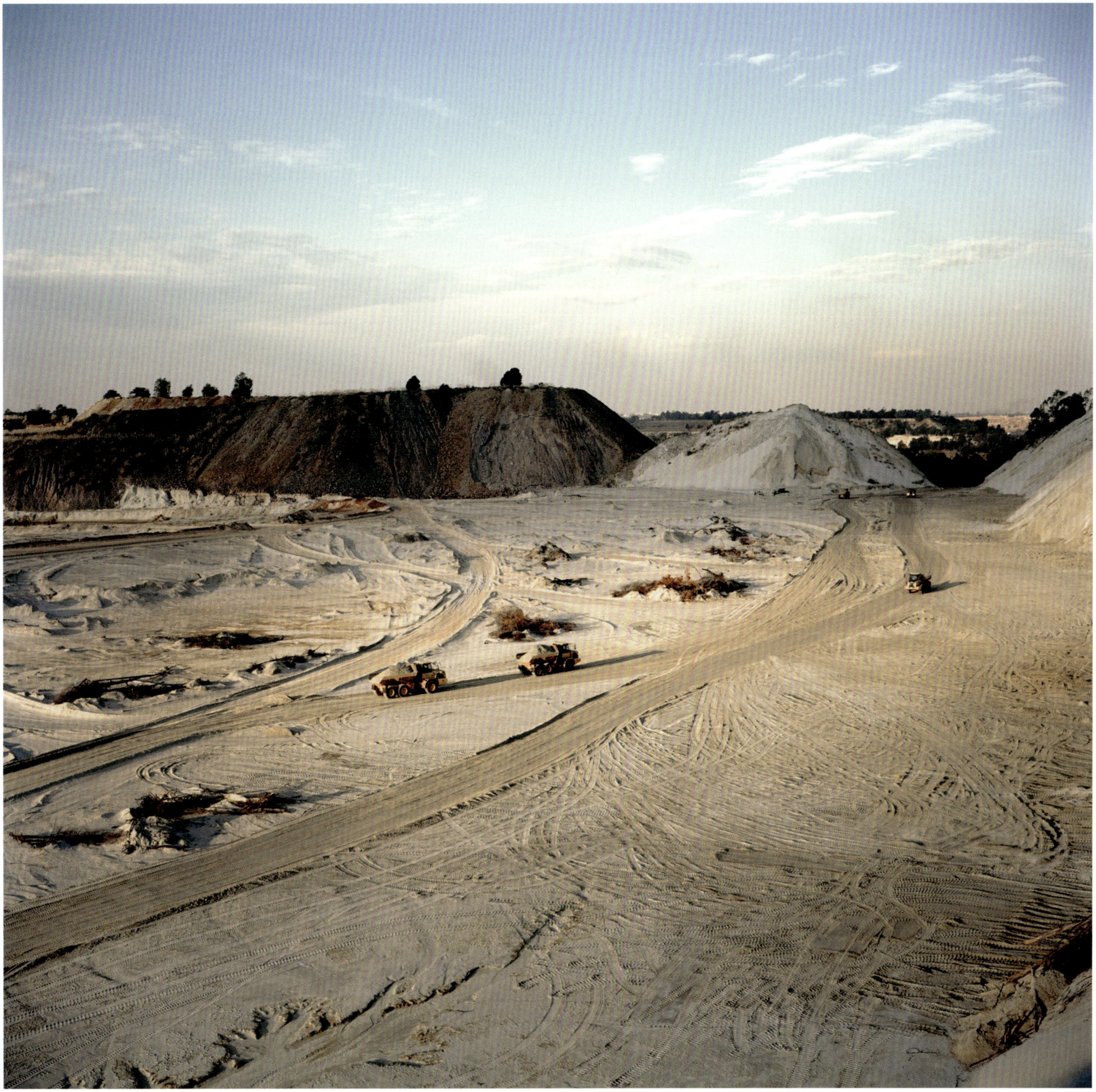

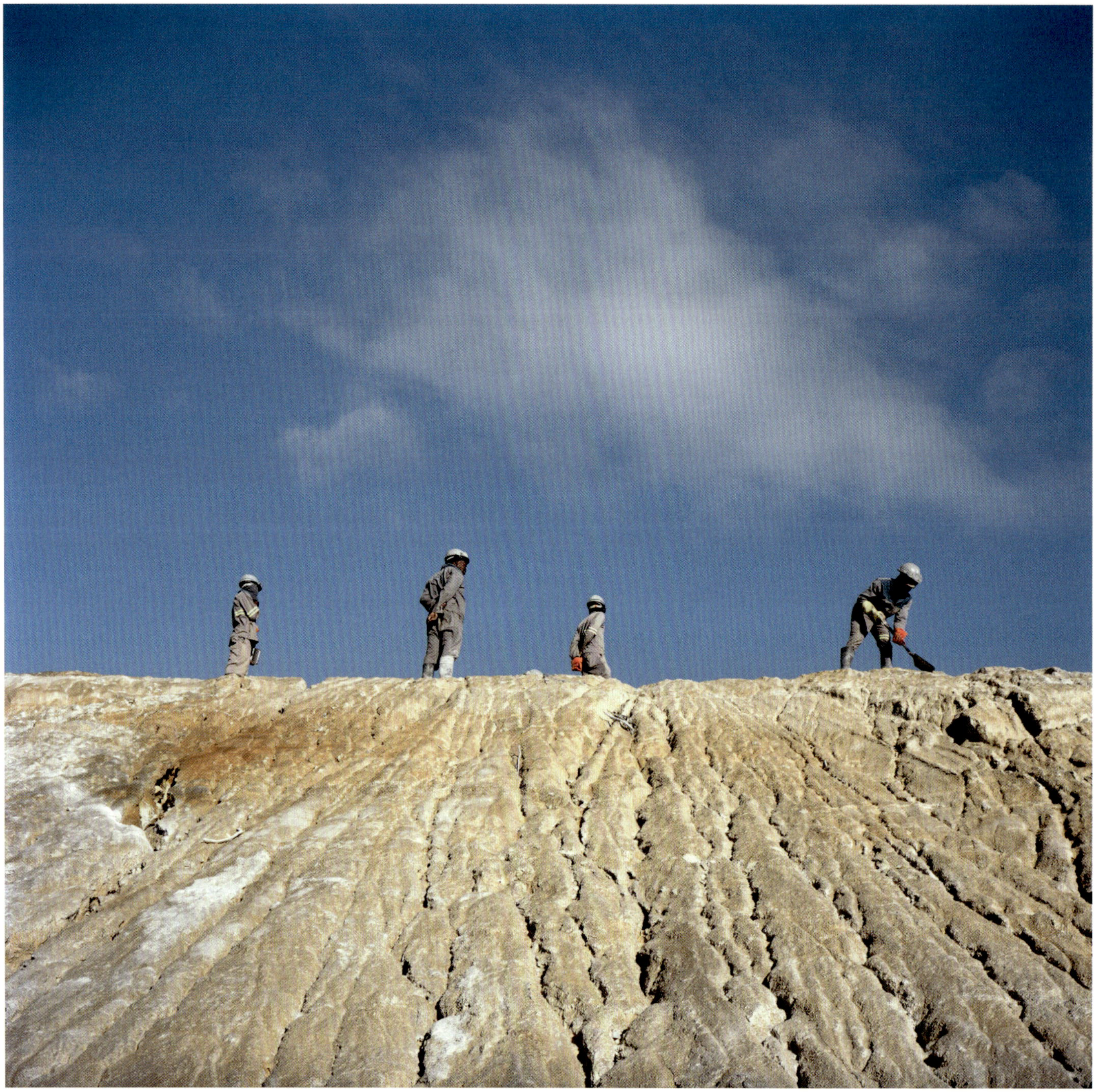

REPAIRING THE DUMP
Krugersdorp, Johannesburg, 2013

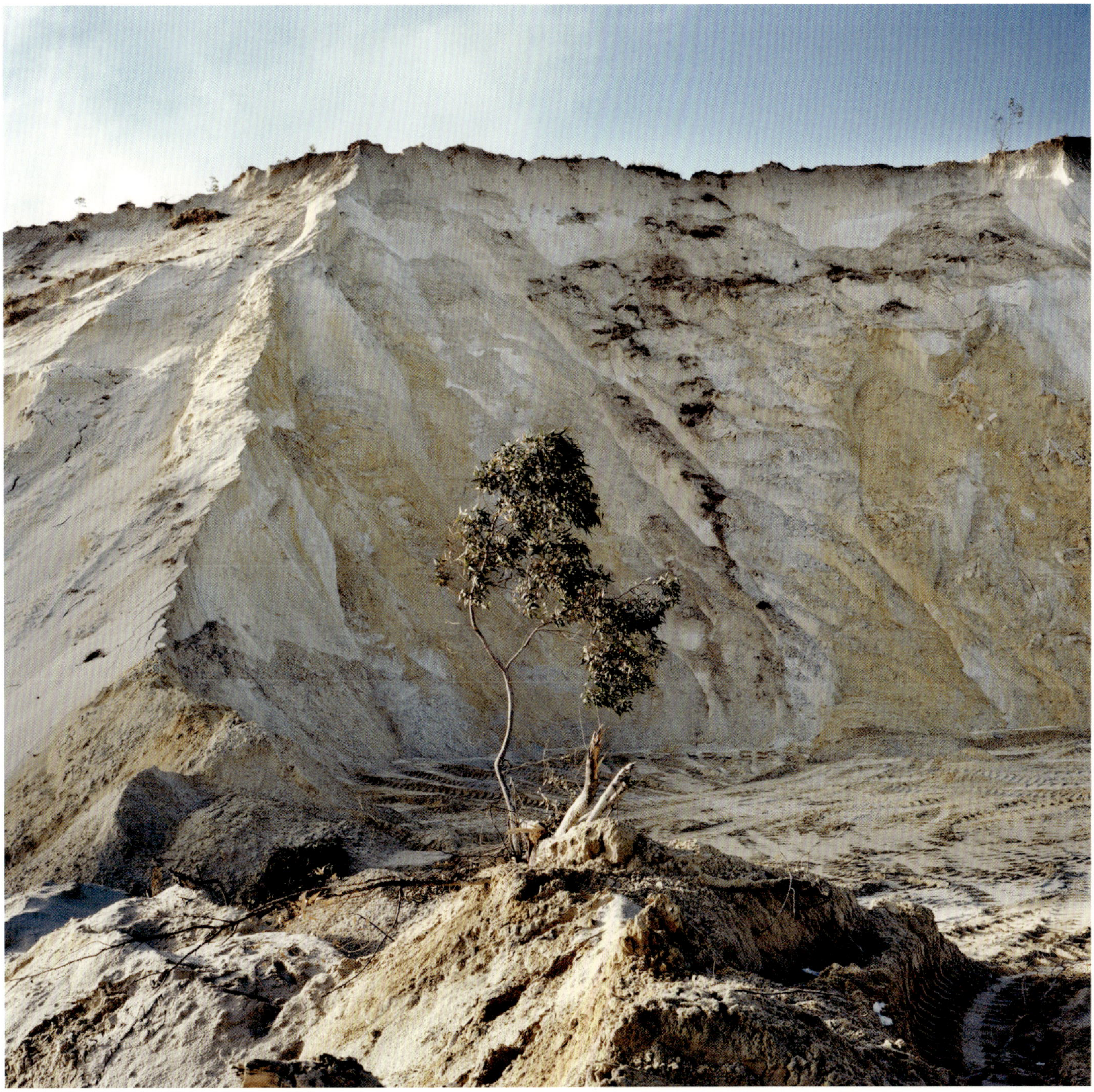

BROKEN TREE
Randfontein, Johannesburg, 2013

ISRAEL MOSALA, ENVIRONMENTALIST
Delmore, Johannesburg, 2012

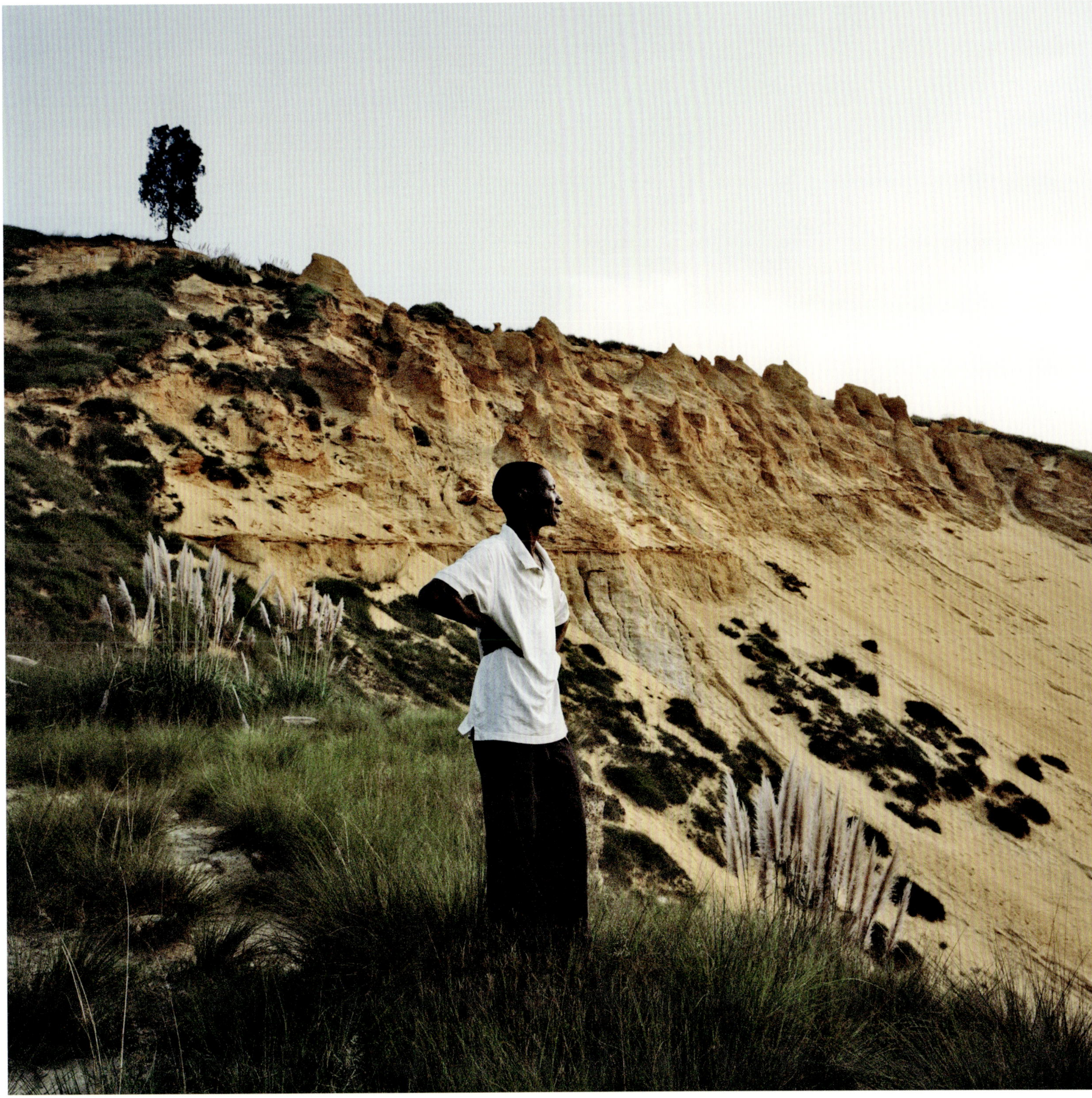

POLLUTED WATER CHANNEL
Driefontein, Johannesburg, 2012

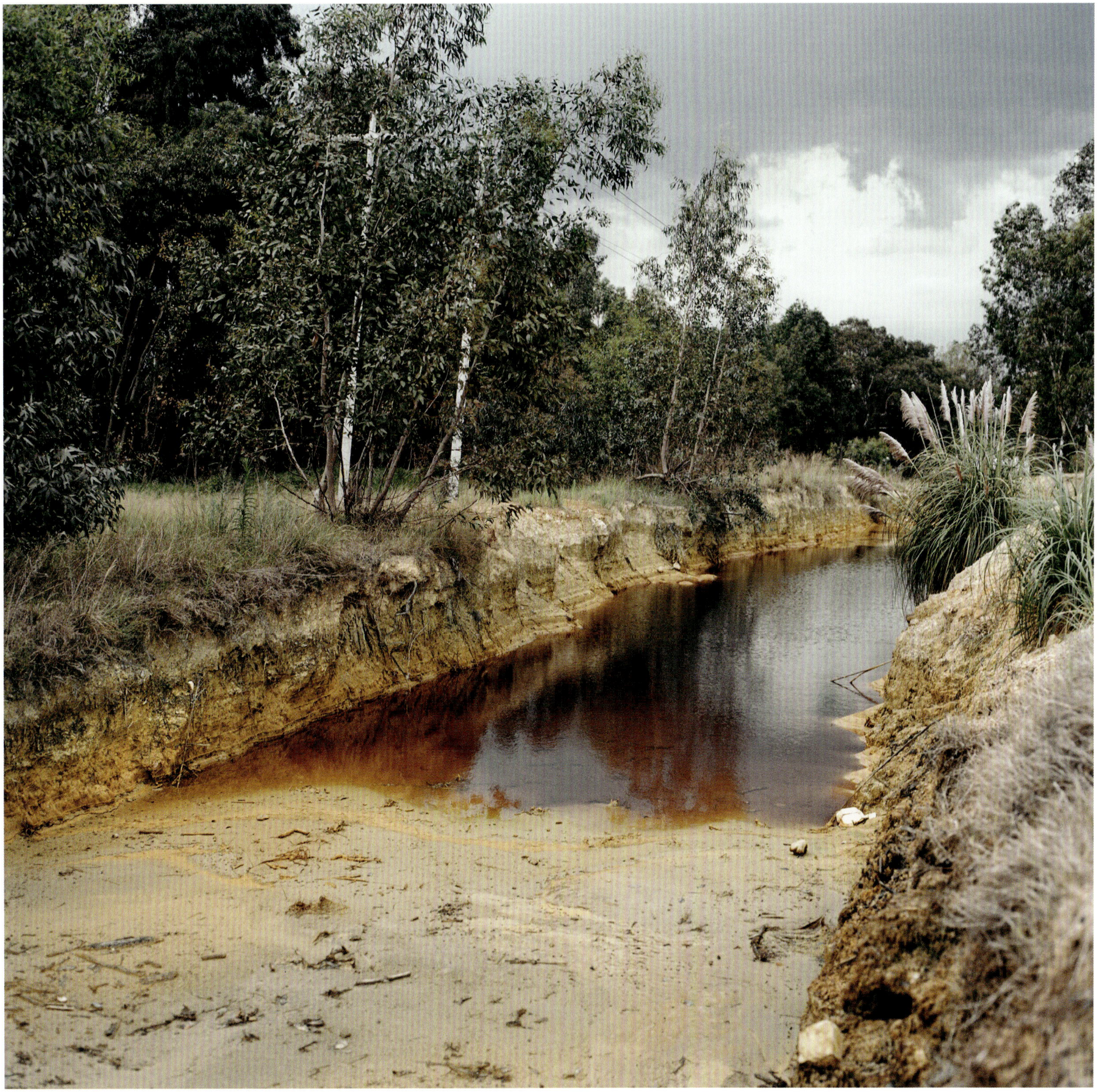

FINAL WEEKS OF A MINE DUMP
Boksburg, Johannesburg, 2010

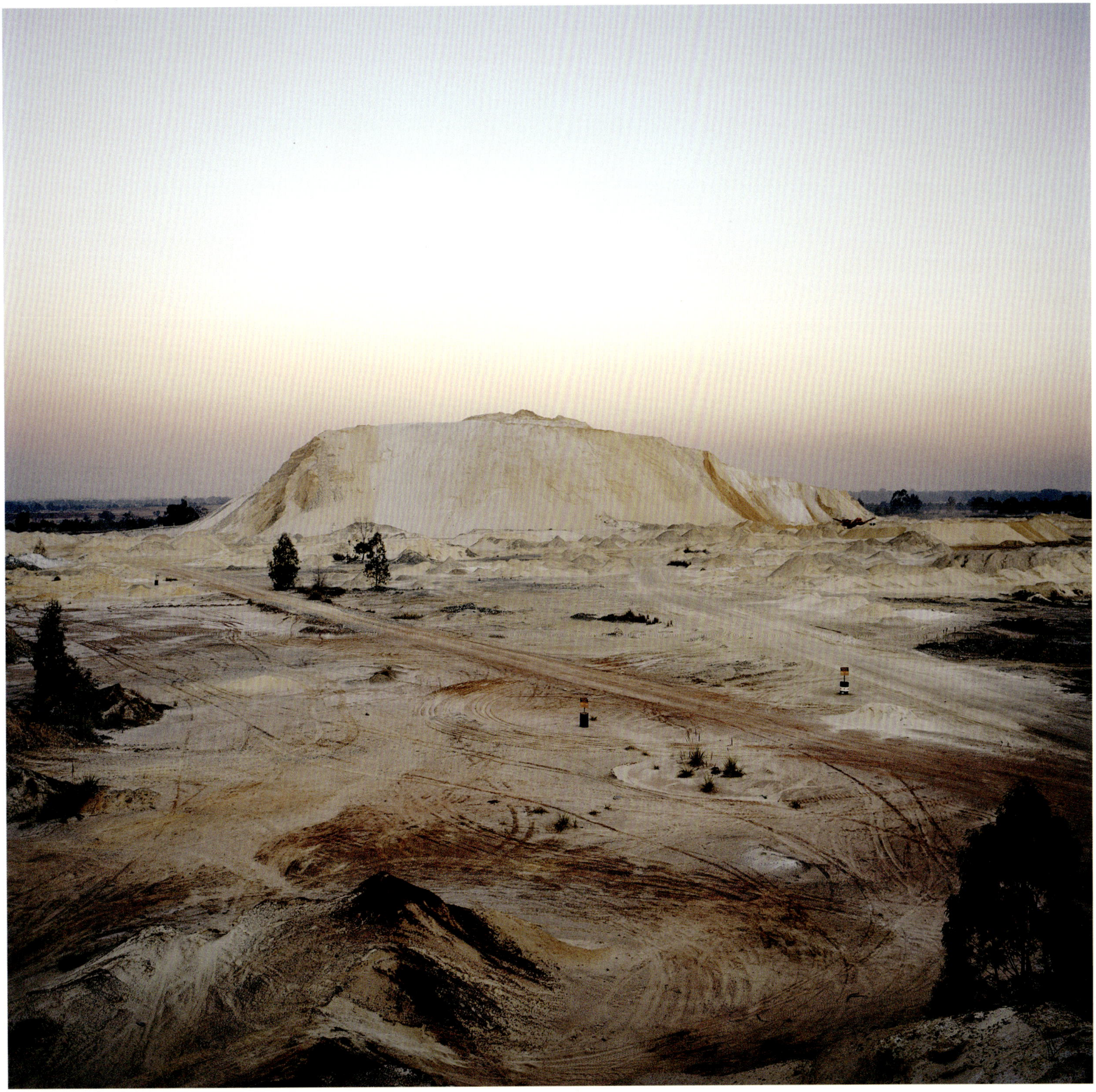

ALL THAT GLISTERS

JULIAN RODRIGUEZ

This is a highly charged landscape. Had gold not been discovered deep below Witwatersrand one hundred and thirty years ago it could still have been just farmland.

The gold might have faded, but Johannesburg was on a footballing high in 2010, the sounds of the vuvuzela loud, when Jason Larkin first encountered the city. What immediately struck Larkin was the almost surreal appearance of the mine dumps – tall, bleached, white and yellow hillocks of crushed rock, pockmarking the city and its environs – incongruous in a land famed for its red-brown soil - and simply structured in a form alien to nature. Largely overlooked by Joburgers, despite their deadly toxicity, these mine dumps are home, a way of life and means of survival for many. Then came the realisation for Larkin of what these dumps represented. A kind of memento mori of the hopes and failures, the booms and busts, the catapulting of Johannesburg from a few scattered farmsteads of the 1890s to a precision-grid modern city in a short half century, the terrible environmental effects of the mining industry (heavily reliant on mercury and cyanide for extraction) and the hideous mistreatment and exploitation of fellow human beings that has always accompanied the gold mining here, and frankly mining all over the world. A history in which Johannesburg rides high, all built on an underbelly of seemingly inexhaustible gold.

Sixty years earlier, following the election of the White Nationalists in 1948, after five hundred years of colonisation-led oppression, the tentacles of apartheid were fully battened down.

Life was always going to be tough, dangerous and unhealthy for the million or so black mineworkers on whose backs Johannesburg's gold mining wealth was built, but in the 1950s it got a whole lot worse as mankind sadly sank to some of its lowest levels of intolerance - rationalised around a constructed and shameful sense of difference. Political activity by Blacks became treasonable, bands worn by Johannesburg gold miners signaled the requirement to distance themselves from their white 'managers' and, in a counter response to apartheid, some of the greatest photography of the twentieth century was produced out of the crowded, buzzing Johannesburg office of a magazine called *Drum*. Staff at *Drum*, at the time a rare multi-racial group of journalists and photojournalists, were busy producing a picture-led magazine, albeit styled on European designs, but squarely targeted at the non-white population of South Africa and keen to bring to light apartheid's tyranny.

The magazine, under the photographic leadership of German-born photographer Jürgen Schadeberg, helped shape brilliant black photographers, a profession unthinkable at the time for non-whites. Schadeberg's protégés included Bob Gosani, Ernest Cole and Peter Magubane; the latter with a camera sometimes disguised in a hollowed-out loaf of bread, fearlessly covered anti-apartheid demonstrations, the 1956 treason trials in Johannesburg and conditions for mine workers in the compounds they were bound to. Later, on the 21st March 1960, he photographed the massacre at Sharpeville on the edge of the city, where the deaths were the result of demonstrations about the Pass Laws – legislation that restricted the movement of Blacks generally and particularly in 'white areas'. Magubane spent in total 586 days in solitary confinement – evidence of the impact his work had on the Nationalist authorities. His inventiveness in disguising his camera was akin to techniques employed by Roman Vishniac, who provoked by the cumulative radicalisation of policy towards Jews twenty years earlier in Eastern Europe, surreptitiously recorded the ghettos for posterity at great personal risk. Like Magubane, Ernest Cole also produced an exposé within the compounds of mines revealing to a shocked world the concrete slabs miners were forced to sleep on and the shameful degrading 'processing' and medical inspections 'tribesman' recruits were subjected to. Cole's book *House of Bondage* (Ridge Press, New York, 1967), a compendium of the horrific excesses of apartheid, remains one of the great photographic books of all time, published after Cole, with print box under arm, fled to New York. *House of Bondage* was to South Africa what Philip Jones Griffiths's *Vietnam Inc.* was to the US – both palpably and powerfully galvanizing public opinion and shaming respective governments. Through *Drum* issue-based photography had become synonymous with political struggle and in its heyday the magazine reached a circulation of over 90,000, despite costing a significant chunk of any black reader's weekly income.

The turmoil of South Africa's recent history has prompted other great photographic work. Magnum photographer Ian Berry, perhaps finding it difficult personally to come to terms with the complexities of South African race politics nevertheless produced very important work: from the riots in Soweto to the building of downtown contemporary Johannesburg, his powerful photographs touched a deep nerve. After *Drum* Paul Weinberg and David Goldblatt continued to bring important investigative stories out into international light and in 1983 helped bring twenty-one concerned photographers, both white and black and many previously not known outside their immediate communities and townships, into focus. Later work by Magnum photographer Stuart Franklin drives home some of the artificial peculiarities of Johannesburg which, propelled by the Gold rush, also mushroomed miles from the original port city grown around its geographical proximity to sea and river.

More recently Roger Ballen's complex and challenging photographs, depicting South Africa's 'poor whites', force us to re-evaluate received historic narratives. Ballen's work also requires the viewer to reflect on the motivations of photographers who seek out difficult-to-look at subjects – a trend started back in 1934 by Lisette Model, Diane Arbus's teacher. (So "upsetting" were some Arbus's portraits that the Museum of Modern Art, New York, had to employ somebody specifically to wipe the spit from the glass frames when her work was exhibited there in 1967.)

In Jason Larkin's photography there is thankfully little artifice to decipher. Part of the power of his recent work lies in Larkin's approach which is, very consciously, less visually dramatic and theatrical, than some of his earlier projects. In his Johannesburg photographs, as with his work *Cairo Divided*, the photographer is once again in deadpan mode; the hard-edged forms and framing perhaps a determined attempt to rid the work of any spurious glamour. But through this extracting and yet also quieter approach, we are drawn into the hopes and fears of the people who live by and on these toxic dumps; and here, echoing the work of US landscape photographer Richard Misrach, Larkin's vision is always carried along by a lyrical sense of space and light. Larkin's approach is also reminiscent of the later work of Mark Power - such as Power's brilliant, if a little subjective, look at the new Poland, *The Sound of Two Songs*. Importantly, for this writer, colour offers little distraction in Larkin's work, a result never easy to achieve technically. But what is most refreshing is Larkin's commitment to unearthing a complex South African story – one resonant of editorial discussions in the *Drum* office in the 1950s. Because of the serious health threat posed by the underlying toxic water table, this story has the potential to be as damaging to Johannesburg as anything this great city has had to endure previously.

Larkin's understanding of the historic significance of the mine dumps grew as he researched Johannesburg's mining history. He began to visit the city's dumps, listened to the people who lived on and around them and started to come to grips with the dump's inherent toxicity and the re-evaluation of their worth as some are re-mined for any left-over gold residues. Despite twenty years of democratisation and reconciliation since apartheid ended, the economic lot of most South Africans has remained unchanged: unemployment and accompanying crime are high - whereas the highly developed financial services industry, much of it rooted in mining, has continued to flourish. Despite disappointment and these enduring strains and race-based tensions, Larkin's forays to the mine dumps were tension-free and the people he encountered readily opened up with their stories.

Continuing to be fascinated by these largely lawless dump sites, in 2011 Larkin went to live in Johannesburg, photographing and conducting interviews and, through official channels got access to the secure compounds where re-mining on an industrial scale is taking place. The trick for an 'outsider' photographer, is in the ability to extract the significance of place – simultaneously disadvantaged and abetted by a lack of familiarity - but still do justice to place and history. Larkin's inclusion within these pages of re-visited archival photographs and a wholly appropriate airing of the mining language of Fanagalo - a fascinating lingua franca of the mining industry used to bridge the language gap which accompanied the influx of mining labour from all over Africa - bring out both the collector and documentarian in Larkin and demonstrate his commitment to the photographic book as a serious interface of contemporary events in the service of history.

Larkin's photographic approach - a careful blending of landscapes and documentary portraits - is detailed and respectful and, as a result, carries some essential truths about Johannesburg – new and old. He firstly avoids slipping into the personalised and introverted approach as exemplified by William Eggleston and secondly circumvents the over-procedural approach championed by the Düsseldorf School (exponents such as Andreas Gursky and later Simon Norfolk in the UK). Instead Larkin is able to be more mobile and offer a nuanced and objective approach rooted in pressing social issues. In the latter respect it would seem Larkin takes heart from Guy Tillim's sobering reality check

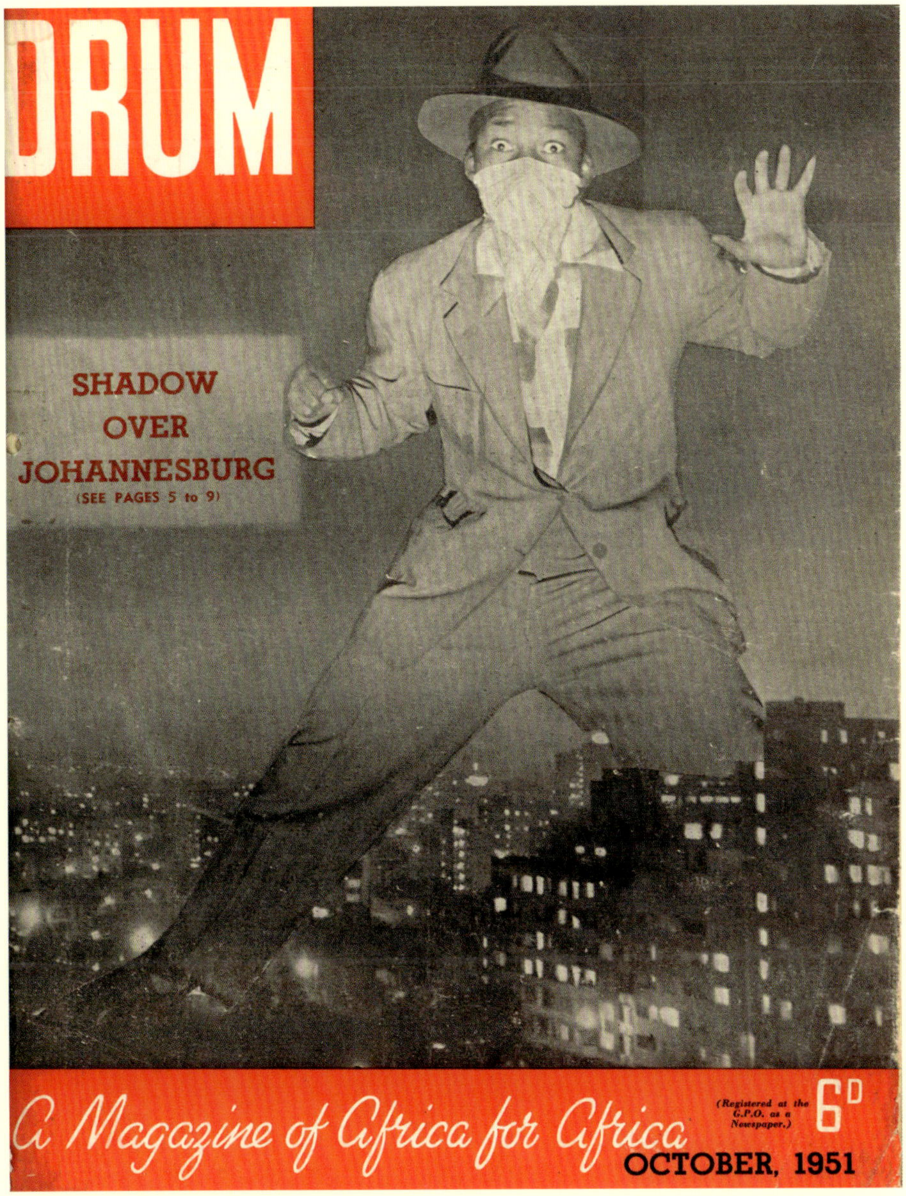

SHADOW OVER
JOHANNESBURG
October 1951

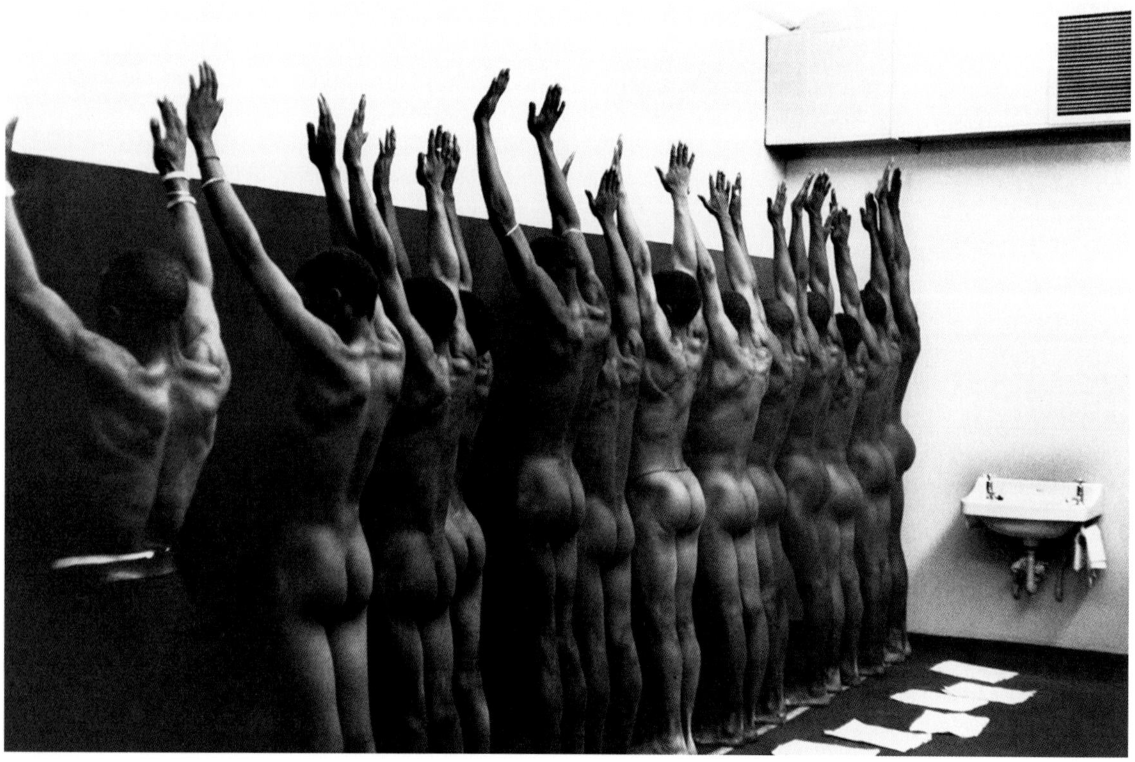

DURING GROUP MEDICAL
EXAMINATION THE NUDE
MEN ARE HERDED THROUGH
A STRING OF DOCTORS' OFFICES.

on South Africa's progress post-apartheid in Johannesburg at the turn of the new millennium. Perhaps he feels affinity too with the subtle but decisive approach to politics evident in Paul Graham's book *Troubled Land: The Social Landscape of Northern Ireland* which relied on almost subliminal details to carry the greater story of the troubles. By not choosing to focus overly on individuals or individual plights, Larkin does reach the bigger picture: unearthing the complexities of Johannesburg's rush to the world stage; identifying the vulnerability and ingenuity that often accompany poverty; measuring the impact of toxic mining in Witwatersrand – all aspects that seep through this larger body of work. Larkin's skill is also in his ability both to quicken and slow the pace at which the story unfolds - pulling us in and out of the landscape and creating a kind of new topography of Johannesburg, definitively pegged to the two hundred and sixty or so dumps within the greater city.

Two girls enjoy a chat on a street corner, nothing much happening, until you consider the toxic dust brought on the winds that sometimes disperses over their daily life close to the dumps in Soweto. The tensions derived from 'coloured' and 'black' historic segregation are also still palpable in Larkin's document. There is edginess in the photograph of Daniel and his hunting dogs, and hope and despair in the Zimbabweans illegally panning for gold in toxic pools. Here, high-powered chemically-enhanced industrial extraction of gold sits cheek by jowl with the age-old methods of individual prospectors, unlikely to squeeze much from the left-over piles. Occasionally, having lulled the reader with the sheer beauty of some of these dunes in soft evening light, Larkin then gives it to us straight – through for instance the fetid orange lake, what the scientists euphemistically call Acid Mine Drainage, the embodiment of the unwelcome detritus of one hundred and ten years of mining and the probable future environmental disaster waiting in Johannesburg's bowels.

With mining essentially more or less over, financial services alongside tourism are now essential to keep this city moving. Here, as in cities and towns the world over the cultural tourism industry reduces and sanitises the past for visitors' easy consumption. However given the unforgettable and still gaping wounds of the recent history of exploitation in South Africa, it is much more difficult to feel at ease with this commodification of history here. So called 'Eco Tours' include "walking in the footsteps of Mandela"; a whistle-stop tour of Johannesburg's leading attraction *The Apartheid Museum*; a visit to Johannesburg's *Chamber of Mines* ("working together for a sustainable future since 1889"); a tour of the historic Magaliesberg: Blaauwbank Gold Mine and 'resort' through to an insight into today's reprocessing of Joburg mine dumps by *Crown Mines*.

Set against this outpouring of tourist-focused, packaged history, Larkin's photographic contribution adds depth, detail and a new perspective on the meteoric and painful rise of South Africa's City of Gold.

FANAGALO

p35 "Don't go into the stope until the bossboy tells you to"

p62 "I can't understand how he fell under the truck.
He is an experienced shoveling boy"

Phrases sourced directly from the mining chapter of 'Dictionary, phrase-book and grammar of Fanagalo, the lingua franca of Southern Africa' J. D. Bold, (Johannesburg : Keartland, 1964). This two-way dictionary, first published in 1951, had multiple editions and was printed into the 1980's in South Africa.

Fanagalo is pidgin form of Zulu and Xhosa with adaptations of modern terms from English and Afrikaans. It was developed in South Africa by colonialists in the early 1900's so as to communicate with local servants. The mining industry, whose workers spoke a vast range of mother tongues from across Southern Africa, soon adopted it as the lingua franca. Only fifteen hours of instruction was considered sufficient for an initiate to become reasonably fluent. The name "Fanagalo" comes from strung-together Nguni forms fana-ga-lo meaning "like + of + that" and has the meaning "do it like this", reflecting its use as a language of instruction

Fanagalo's use spread as far as mining operations in the Congo and Zambia and is a rare example of a pidgin based on an indigenous language rather than on the language of a colonising or trading power. Today it is still used in some mines across South Africa, even after attempts by mining companies to stop its use.

EXTENDED CAPTIONS

P5 New road connecting Soweto to Johannesburg built in the run-up to the World Cup, pictured here after the games, as yet unopened.

P13 A Metrorail train to Park Station, central Johannesburg. Pylons, train lines and other vital infrastructure crisscross mine dumps across Johannesburg.

P15 RDP refers to Rural Development Program, initiated by Nelson Mandela and the ANC to provide housing for millions of South Africans after apartheid rule. These houses are built in contravention of the Mine Health & Safety Act, which stipulates that no mining activity can take place within 100 meters of residential dwellings.

P17 Coloured is used in South Africa to refer to people of mixed-race parentage rather than, as elsewhere, to refer to African peoples and their descendants. Under apartheid it was imposed as an official racial designation. In modern use the term is not generally considered offensive or derogatory.

P18 DRDGold owned dumps that surround the communities of Diepkloof. They are still in use and can grow in height by around 10metres a year from deposits of new mine waste.

P25 Sporting facilities originally built for a mining college established in the early 1900's by Crown Mines, now run by Central Johannesburg College as a music and visual arts college.

P26 Numerous evangelical Christian groups within South Africa have no formal churches or premises to use. Mike prefers to pray on the mine dump after finishing his shift in a nearby factory.

P29 The outdoor cinema was built on a mine dump in 1956. DRDGold closed the cinema to recover residual gold left in the dump. A reported ten million dollar's worth of gold was recovered over five years.

P39 Acid Mine Drainage (AMD) refers to the outflow of acidic water from mining sites and is created through the oxidation of metal sulfides from exposed and milled rock.

P51 Nicolle, a member of the Zion Christian Church, is encouraged by his order to consume fresh water at least once a week. This water runs directly from the mine dump in the background and is reportedly contaminated with AMD.

P67 Children from the community of Davidsonville play with sand that's been washed down from the mine dump located only meters away from their family homes.

P71 Pressurised water is used to break down the dump into a sludge for easier piping to the processing plant, where most of the remaining gold content is then removed.

P77 Dump 20 was listed in the Guinness Book of Records as the largest man-made heap in the world. It took sixty years to create and will take around twelve years to remove and re-mine.

Jason Larkin, (1979) is a British photographer recognised for his desire to pursue subjects on the periphery of current affairs. Following his study of photography at Falmouth University and Westminster University, Larkin worked as a documentary photographer across the Middle East and Africa, with his work published worldwide. He was recently awarded the PDN Arnold Newman New Portraiture Award and the Ithuba Arts Fund. His freely-distributed publication, *Cairo Divided,* was nominated for the Deutsche Börse and Prix Pictet photography awards. Exhibitions include the Brighton Photo Biennial, Flowers Gallery, London, and a solo show at Farnsworth Art Museum, USA.

Julian Rodriguez is Head of Media at Falmouth University. His book, *Roman Vishniac*, was published in 1998 and he has written extensively on both the influence of photography and its impact as historic documentation for the British Journal of Photography, London. He has been Chair of Photoworks, Brighton, England since 2010.

Mara Kardas-Nelson is a South African-based journalist focusing on health, human rights, the environment, social justice, and local culture. She is currently housed at the Mail & Guardian's Centre for Health Journalism, having previously worked with the Globe & Mail, rabble.ca, and researcher and writer Naomi Klein. Her work has appeared in North American, European, and African publications.

PHOTOGRAPHS: Jason Larkin

TEXTS: Mara Kardas-Nelson and Julian Rodriguez

ART DIRECTION & DESIGN: Realise Creative

IMAGE REPRODUCTION: Erwin Schulz & Lightfarm

ILLUSTRATIONS:
Ester Svensson from original photographs of rock drill operators working in Johannesburg gold mines. c. 1900's (Unknown) Sourced from Museum Africa.

FRONT & END IMAGES:
Details from gold mine shaft photographs. c. 1910 (Unknown) © Museum Africa

ACKNOWLEDGMENTS:
Akona Kenqu, Benedicte Kurzen, Celine Smith, Chris Littlewood, Claire Gaze, Claire Johnson, Dick Plaistowe, Eddie Milne, Elles van Gelder, Ethel Zucker, Fourthwall Books, Francesca Beard, Francesca Sears, Harry Hardie, Isreal Mosala, Jack Shenker, James Greenhow, Jonathan Torgovnik, Kate Edwards, Lin Nelson, Mal Stone, Maja Daniels, Mariette Liefferink, Max Houghton, Olivier Laurent, Panos Pictures, Patrick Waterhouse, Peter Kardas, Shyam Goswami, Suzanne Bull, Ithuba Arts Fund, Zed Nelson.

Published with kind support from Falmouth University

FALMOUTH
UNIVERSITY

Kehrer Heidelberg Berlin
www.kehrerverlag.com

ISBN 978-3-86828-416-4
Printed in Germany

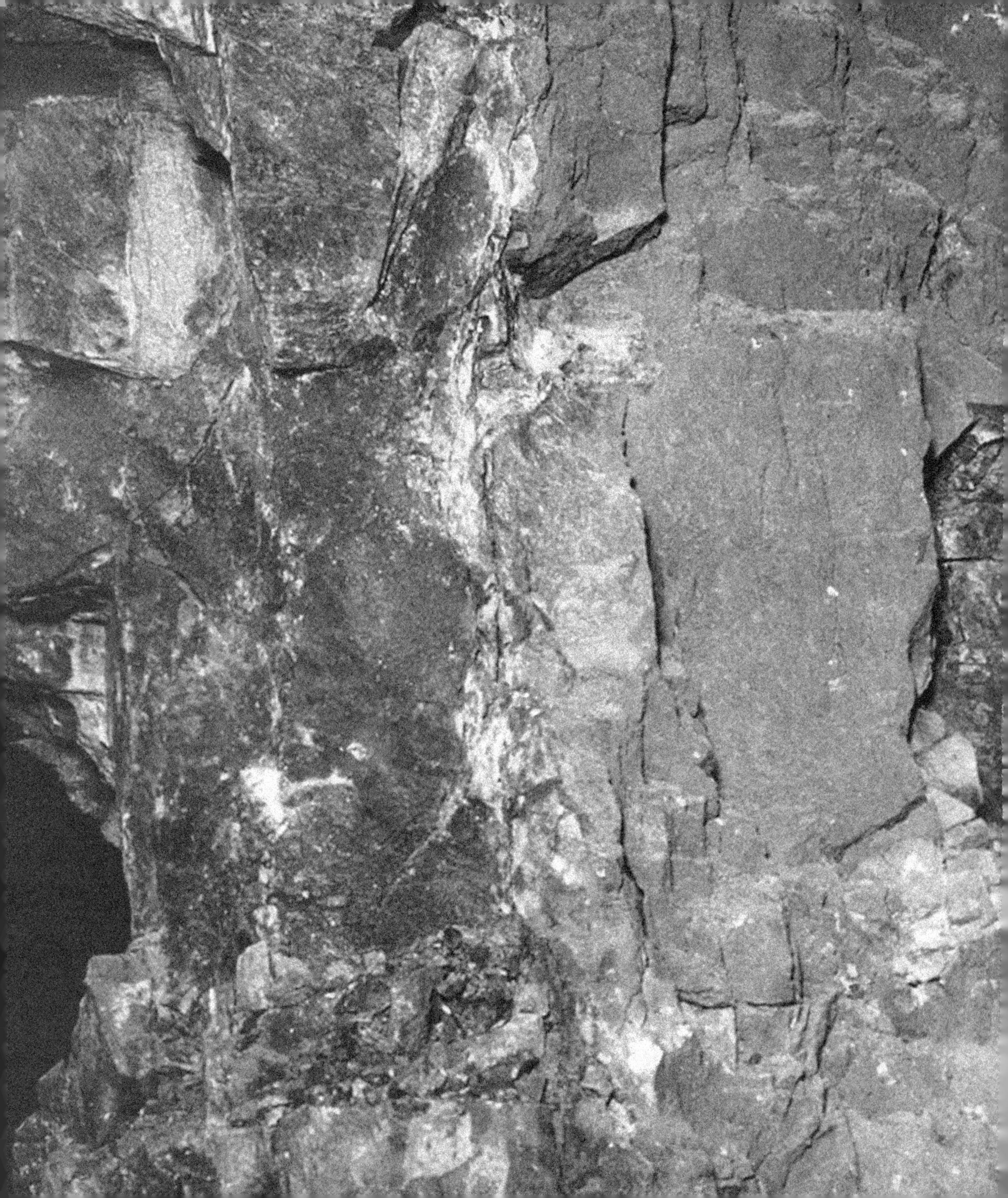